AF522001

INTRODUCTION TO LOGISTICS MANAGEMENT

Carl M. Guelzo

INTRODUCTION TO LOGISTICS MANAGEMENT

A Reston Book
Prentice-Hall
Englewood Cliffs, New Jersey 07632

Library of Congress Cataloging-in-Publication Data

Guelzo, Carl M.
Introduction to logistics management.

"A Reston book."
1. Business logistics—Management. I. Title.
HD38.5.G84 1986 658.7 85–18460
ISBN 0–8359–3221–4

Editorial/production supervision and interior design by Norma M. Karlin

A Reston Book
Published by Prentice-Hall
A Division of Simon & Schuster, Inc.
Englewood Cliffs, NJ 07632

10 9 8 7 6 5 4 3 2 1

Printed in the United States of America

CONTENTS

PREFACE

This is the place in every book, I am told, where authors are supposed to convince readers that the writing is not motivated by sheer greed. Unless the writer is a Samuelson with his economics textbook or a Meigs in accounting, the motivation for writing a book certainly must lie in some other direction. For this book, that direction is communication with a much larger audience than can be attained with a technical journal article or in a single classroom. The message is simple enough: Tell all within sight and hearing of the growing importance of the management of the logistical process.

The practice of logistics is as old as the first hunting expedition in prehistory; the logistical historical stream continued through the travels of Bronze Age salesmen, who peddled the products of their home regions in distant lands, to the traveling fairs that were so important to the revival of trade in medieval times. Logistics, as a career field, is relatively new in point of historical time, stimulated if not born in the massive movements of personnel and materiel during World War II.

In view of the growing importance of logistics and management, a need exists for a volume intended to come on the market at the introductory level without trying to touch all bases in a single, weighty, encyclopedic tome. In consequence, this book is intended for a multitiered audience:

- Those without any background of any kind in logistics who need a basic working knowledge of the field either through personal interest or as part of a work-related requirement.
- Those who need some information about the field as input into a decision about what career to enter.
- Those with a smattering of knowledge about logistics and the need for more either because an academic program or the job requires a more extensive background.
- Those who are already in a managerial position and are about to have some aspect of logistics management added to their job responsibilities and need to know what logistics is all about and what logisticians do.

The material contained in this book is self-sustaining and requires no particular background in other technical areas. The expectation is, however, that further work in logistics management will include material in such allied technical fields as purchasing and transportation, which will be mastered concurrently or shortly after finishing this book. The mathematical background required is minimal; but where mathematics is needed to present a topic more clearly, mathematics is used without apology or hesitation. Numbers are so much a part of management generally and logistics specifically that some mastery of mathematics is essential.

The set of questions and problems provided at the end of each chapter is intended to test the extent to which the chapter objectives have been achieved. The problems, of necessity, are numerical in nature; but the discussion questions, with few exceptions, are in the essay format. If logistics managers need mathematics, they also need even more a mastery of communication skills. This book is not structured to teach communication, but it can give the readers some much needed practice in expressing themselves in the written language. Since communication is so important in management generally, I would also add to the communication need the requirement to learn how to listen. Listening is an art much decayed in our society, and I regret deeply that this work is unable to give more much needed practice in that area.

Working through the end-of-chapter exercises is highly recommended, since the allowed practice of turning back to reread pertinent portions of the chapter will tend to reinforce learning. Answers are also provided to give immediate feedback, but those questions requiring interpretive, essay-type answers are provided solutions that can best be described as tentative. Individuals bring to their study of logistics vastly different experiential backgrounds. The points of view from which these individuals regard the questions can be used to enlighten and expand the given answers far beyond the suggested responses. All this activity, of course, is intended to enhance the use of this book as a learning experience.

A glossary of the technical terms used in the book is provided, and a list of references is included for those who wish to read further in logistics management. The books listed contain a vast array of further technical material in all aspects of logistics and management. In a field growing as fast and dynamically as logistics, many of the important technical advances, as well as invaluable on-the-job-type hints to make the work a bit easier, are reported in the professional journal literature before being enshrined between hard covers. In recognition of this fact, the reference list contains a section citing some of the major journals in the field.

Perfection is a most elusive goal and I do not entertain the illusion that this book has attained that ideal. I do, though, want this to be most useful, if not the best, book in logistics management, given its modest aspirations to excellence and coverage. In pursuit of that goal, please write me at the following address should you have any comments, suggestions, or expressions of anguished rage:

Dr. Carl M. Guelzo
℅ Prentice-Hall
Englewood Cliffs, NJ 07632

I hope this guide to your travels through logistics management will be informative, with perhaps the added bonus of being enjoyable as well.

ACKNOWLEDGMENTS

If anyone thinks writing a book is a solitary endeavor, let them now prepare to change their mind. A surprising number of people are involved, many never explicitly identified, who make a contribution (e.g., printers and binders, typographers, sales representatives, and production workers). This, then, is for those both identified and unknown. A tip of the hat to Ted Buchholz, formerly of Reston Publishing Company, who conned me into an effort of such length and intensity as I have never experienced before. A salute to all those who who reviewed the manuscript in various stages of development and made more helpful comments and suggestions to improve the work than can be enumerated here.

I did my own typing, so I can thank myself for that, although the proofreader will have other ideas about my typing skills; but to me alone belong all the errors that must certainly still lurk in the text, despite a ferverent wish to blame someone—anyone—else.

INTRODUCTION TO LOGISTICS MANAGEMENT

part 1

THE LOGISTICAL ENVIRONMENT

1

Logistics: Physical Distribution and Materials Management

CHAPTER OBJECTIVES

Upon completion of this chapter, you should be able to:

A. Define some of the most commonly used logistics-related terms.

B. Explain the relationship of logistics to other organizational elements of a company.

C. Differentiate between materials management and physical distribution.

D. Appreciate the importance of logistics in controlling costs.

E. Understand the function of logistics in assisting in the managerial decision-making process.

Look at your television set. No, don't turn it on; just look at it. What a complex bundle of materials, parts, wiring, and who knows what else! Ever think how all that was brought together to appear before you as a TV set? By the miracle known as logistics, that's how. What is this modern combination of magic carpet and Aladdin? A definition of the subject matter of this book would be a good place to start:

> BUSINESS LOGISTICS: All activities required in managing the movement of raw materials to and through production facilities and finished goods to consumers.

Like most definitions, this one both conceals and reveals. To reveal some of the concealment, we might take a page from the book of the economist in describing what business logistics actually does in a market system.

CONTENT OF BUSINESS LOGISTICS

The working of the economic system by which goods and services are supplied to consumers involves four basic market functions: production, distribution, exchange, and consumption. Logistics assists in the efficient performance of each of these functions. *Production* transforms raw materials into finished goods (i.e., creates *form* utility). In so doing, a long and intricate logistical chain is activated to bring the material together in the proper quality and quantity at the right time in support of the productive process. Much of this book is devoted to a description of the management of that logistical chain.

The function of *distribution* places raw materials in the hands of producers and finished goods in the hands of consumers when and where wanted (i.e., creates time and place utility). Transportation comes to mind as a key element in this link of the logistical chain; but getting goods where and when needed involves, as we shall see, much more than just the service of carrying products from here to there.

The function of *exchange* transfers goods from the wholesale to the retail links in the chain, while *consumption* ends the process. All this might sound simple enough; but when the "consumer" is a battalion of infantry crossing the beaches of Normandy early on the morning of June 6, 1944, or a handful of astronauts perched atop a flaming rocket, the sheer enormity of the logistical task can boggle the imagination. The logistical aspects of a brief visit to the presidential retreat at Camp David, Maryland, can be just as problem filled. Apparently, logistics has been woven into the very fabric of modern life. How else could the seemingly insatiable public appetite for automobiles, houses, television and stereo sets, household appliances, and entertainment be satisfied?

We cannot here solve the problems of the world, but we can do something about the logistical problems of business enterprises. At the level of the individual firm, this will be done using techniques appropriate to the two major branches of business logistics: materials management and physical distribution. Which brings us to two more definitions:

> MATERIALS MANAGEMENT: All activities concerned with the management of the flow of raw materials and semifinished goods from the sources of supply to the point of manufacture.
>
> PHYSICAL DISTRIBUTION: All activities concerned with the management of the flow of finished goods to consumers.

Some niceties in the use of the preceding terms should be observed. A "finished good" to one company could be an input to another. For example, tanned leather hides are finished goods for the tanneries, but are raw or semifinished material inputs to the shoe manufacturing company.

What, then, are these various activities?

MATERIALS MANAGEMENT

Materials management includes the service of transportation, inventory management, the acquisition of materials (or purchasing), the storage of materials once acquired, and the handling of the materials while in the process of manufacture. Physical distribution continues the process.

PHYSICAL DISTRIBUTION

Physical distribution also includes transportation, but this time outbound from the plant or storage facility to customers. The management of the finished goods inventory is included here, along with the protective packaging of goods to reduce damage in transit (marketing handles the type of packaging designed to attract purchasers and sell products), and storage and materials handling. Both branches of logistics, then, deal with many of the same functions but from different points of view: Materials management is mostly concerned with the inbound flow of materials, while physical distribution deals primarily with the outbound flow.

Logistics thus bridges the gap between consumer demand and producer supply, although the idea of consumer can include both individual users of a product and the user of raw materials or finished goods produced by someone else. The shoe manufacturer is as much a consumer of tanned leather as the ultimate wearer of the footgear. A producer can include both the gravel pit operator who provides rock to the cement company and the contractor who "produces" paved roads from the gravel and cement.

IMPORTANCE OF BUSINESS LOGISTICS

Prior to World War II, businesses were generally small enough to solve their logistical problems by the personal experience and intuition of their own personnel. Input materials were both cheap and plentiful and the means of hauling everything around readily available. World War II, with problems involving the movement of huge quantities of supplies and large troop units under the most adverse possible conditions, raised logistics from glorified troubleshooting to a distinct technical field. The post-World War II business environment restricted somewhat the range of logistical problems to be solved. The general public, with bank accounts bulging with wartime savings, kept their demand for goods pressing upon the ability of producers to supply products. The economic problems of the 1950s thus centered on a slack economy, a manageable amount of unemployment, and slowly rising costs of doing business. These problems shifted the attention of managers to profitability.

Rising interest rates and the oil-embargo-induced leap in energy bills focused the attention of businesses on the costs of their logistical operations. Suddenly, cost control became an important element in profitability, and logistics took its place in cost control systems. As the years approached the 1980s, plowing through the turbulent times of the Korean and Vietnam wars, the legal and regulatory climate in which business operated became more complex. Shippers discovered that regulatory agencies would hear their testimony as well as that of the carriers and competing manufacturers; computers permitted the application of ever more mathematically intricate quantitative management techniques, and labor costs accelerated so rapidly that the sheer detail required in managing a large enterprise seemed almost overwhelming.

Logistics quickly became involved in every major function of management. Almost any book on management will list these managerial functions; let us here keep life as simple as possible by restricting the list to four.

Planning, for example, involves looking ahead to determine the most efficient (i.e., cost- or profit-effective) way of achieving company objectives. This would require, among other factors, an orderly flow of materials into and out of a company, while at the same time achieving a low-cost operation. You can picture the logistician at work here.

Organizing brings together tasks and a mixture of human and material resources to achieve company objectives. This would require an orderly flow of materials within the company itself and could easily involve purchasing functions. More logisticians at work here.

Directing includes both giving orders and ensuring that sufficient resources are available to support execution of the orders. More jobs for logisticians.

Finally, *coordinating* ensures that the component elements of the company work together in an efficient, low-cost way. This, too, involves the logistician in the overall company system.

LOGISTICS AND THE COMPANY SYSTEM

As corporate enterprises become larger and more complex, an easy error into which to fall is forgetting that a business is frequently greater than the sum of its organizational elements. Responsibilities become fragmented into smaller and smaller specialties. The members of these small, concentrated units are then judged on how well their fragment performs, rather than how much of a contribution the fragment makes to the total operation of the firm.

For example, customer service gets assigned to sales; production goes to manufacturing, and inventory control gets split between purchasing and warehousing. Almost instantly the problem of the trade-off arises: Sales wants the best possible customer service, but this noble aspiration requires high inventories for the desired instant response to customer orders. Production must also be flexible to meet sudden shifts in demand. Thus, if sales gets what it wants, the performance of manufacturing, purchasing, and warehousing deteriorates due to the high cost of carrying large inventories and scheduling irregular production runs in the factory.

What manufacturing really wants is a nice, long, uninterrupted production run. Unfortunately, this, too, requires large inventories, what may be an unacceptable loss of flexibility in meeting changing customer demand, and some delay in filling customer orders that arrive just as a production run is started to refill a depleted finished goods inventory. The result is frequently suboptimization: Each organizational element strives for perfection in its own area of responsibility without regard for the impact on the total company system.

For example, purchasing concentrates on submitting purchase

orders on schedule and tracing late orders, but is not overly concerned with the problems of storage because that function belongs to the warehouse. The formation of a company system involves acceptance of the idea that a cost-efficient operation in one part of the organization does not always mean an efficient operation elsewhere.

Achieving the lowest overall transportation cost may not be compatible with the shortest possible delivery times to customers. The company system requires that component elements of the organization be interconnected to form a single corporate entity, rather than a loose confederation of individual functions. Working with an information system, the company system can measure performance of the individual units and present a total picture to management for analysis.

What is needed is a considerable degree of direct cooperation between all components of the company to ensure that all efforts are directed at achieving the objectives of the firm. This requires that all organizational elements have inputs to plans, even though the plans are the responsibility of production, marketing, purchasing, or engineering.

WHAT CAN LOGISTICS DO FOR MANAGEMENT?

Consider what must be done to put a rock group on tour. Within the organization, logistics can group related functions so that support becomes mutual (e.g., getting the musicians and soloists from one engagement to the next), inventory control (e.g., keeping track of instruments, sheet music, and sound equipment), warehousing (e.g., locating suitable accommodations for all personnel and storage space for the equipment), acquisition (e.g., purchase of new and replacement sheet music plus repairs to the equipment), and materials handling (e.g., loading and unloading people and equipment).

Logistics supports marketing in the functions of customer service (e.g., fending off screaming hordes of groupies and autograph hunters), order processing (e.g., handling the flow of bookings for engagements during the tour), selling and promotional activities (e.g., advertising in local media, personal appearances in advance of the performances, posters and sales of tickets and related souvenir-type products), packaging (e.g., persuading the media that this is really a bunch of sweet kids who are kind to dogs, children, and grandparents), and product mix (e.g., which shall it be among hard rock, punk rock, regular rock, blues, country and western, or whatever in the proper proportion for each concert).

Logistics also supports production in materials handling (e.g., more loading and unloading of buses, trucks, planes, flying carpets, etc.), in-house supply (e.g., getting this cantankerous group of prima donnas from dressing or practice rooms to the stage), plant and storage area locations (e.g., how far from the concert hall should musicians and equipment be placed), acquisition (e.g., obtaining food and drink—exact nature unspecified—at the appropriate hours), and a serious attempt to reduce losses due to damage and pilferage (e.g., keeping the audience from grabbing bits of instruments, clothing, hair, contact lenses, and whatever else can be separated from the performers).

CONTRIBUTION TO INTERNATIONAL TRADE

A company with an efficient system of logistics also has a contribution to make external to the organization. On a geographical basis, not every region is similarly endowed with productive resources. Logistics permits each area to concentrate on doing those things or producing those products that can be done best to improve the welfare of everyone. South Korea, for example, is not richly endowed with natural resources, but is blessed with a literate population capable of an enormous amount of hard work and persistence. The Republic of Korea, thus, does not attempt to export raw materials to the world, but rather concentrates on the production of goods that require much expert labor: textile, electronic goods of various kinds, and small appliances, to name but a few. The rest of the world gets, as a result, first-rate quality shirts at a price much lower than if other nations tried to become self-sufficient in textile goods.

Australia, as another example, is short on population but long on vacant land area. The result is the export of meat products, which need much land but relatively little labor or machinery. In consequence, the rest of the world gets excellent Australian mutton and fine wool, which would be obtained at a much higher cost without the Australian specialization. The United States is good at mechanical products and, as a result, even the Arab world prefers American medical equipment.

Without logistics, each area of the world would be forced into self-sufficiency with a great deal of high-cost, inefficient production. Logistics permits the conduct of international trade, and thus more goods are available to everyone, contributing to a generally higher standard of living. How much better food began to taste in medieval Europe when a rudimentary trade served up spices in return for local products. International trade to the extent enjoyed in the modern world is made possible in large measure by a well-developed system of logistics.

What more could be expected of an area of human endeavor that contributes so powerfully to the well-being of individuals and to the organizations in which they work. This is especially significant when considering that logistical activities already absorb up to 15 percent of the entire gross national product of the United States. In studying logistics, we are not considering a minor area of human activity, but one that affects almost every aspect of daily personal, public, and commercial life. The variety of activities is similarly broad and easily provides the basic materials from which highly satisfying careers can be made.

OBJECTIVES ACHIEVEMENT CHECKUP

The following exercises will help you in assessing the extent to which you have achieved the objectives for this chapter. Where essay-type answers are required, the more you are able to condense your knowledge into a short paragraph of a few sentences the better. Looking back as you work these exercises and rereading parts of the chapter are permissible; but do not look ahead to the answers until you have done your best with the questions.

1. Match the logistics-related terms in Column I with their definitions in Column II.

Column I	*Column II*	
A. Production	(a) Managing the movement of raw materials and finished goods	(a) ___
B. Organizing	(b) Managing the flow of materials into a company	(b) ___
C. Coordinating	(c) Managing the flow of finished goods to consumers	(c) ___
D. Exchange	(d) Creates form utility	(d) ___
E. Business logistics	(e) Finding the most efficient way of achieving company objectives	(e) ___
F. Directing	(f) Places products in the hands of consumers	(f) ___
G. Distribution	(g) Brings tasks and resources together	(g) ___
H. Physical distribution	(h) Includes inbound transportation and inventory management	(h) ___
I. Planning	(i) Giving orders	(i) ___
J. Materials management	(j) Transferring goods from wholesale to retail channels	(j) ___

2. Would it be correct to say that logistics is an organizational element of a company in the same sense as engineering or accounting? Explain.
3. What functions are common to both materials management and physical distribution? What functions are not common?
4. What contribution can logistics make to cost control?
5. What inputs can logistics provide to managers when making decisions?

ANSWERS TO CHECKUP QUESTIONS

1. (a) E (f) G
   (b) J (g) B
   (c) H (h) J
   (d) A (i) F
   (e) I (j) D
2. No. Logistics is a function that ties together other functions performed in an organization. Logistics is not a type of work done in only one unit or department of a firm.
3. a. Common elements: Transportation, inventory management, and storage.
   b. Elements specific to materials management: Inbound transportation, raw materials inventory management, storage of materials to be used in making a product, and the intraplant handling of materials in process.
   c. Elements specific to physical distribution: Outbound transportation, finished goods inventory management, storage of finished goods, protective packaging.
4. Logistics can contribute to cost control through improved efficiency in using the service of transportation and in minimizing the cost of purchased materials, minimizing the size of inventory, and reducing the length of time maintained in storage.
5. Logistics can contribute to the managerial decision-making process by indicating the way an orderly flow of materials can be achieved and thus aiding in deciding on the best way of achieving company plans. In addition, logistics can assist in balancing the desires of departments with the most efficient manner of operating the company.

2

Logistics in Support of Purchasing

CHAPTER OBJECTIVES

Upon completion of this chapter, you should be able to:

A. Explain the purchasing function.
B. Understand the relationship of purchasing to other logistical functions.
C. Describe the placement of purchasing in the organizational structure of a business enterprise.
D. Differentiate between the various types of ordering techniques.
E. Appreciate the place of logistics in the ordering process.

Since logistics is so closely tied into the costs of doing business, why the concern with purchasing? Because over half the value of a finished product is added by the purchased input materials. If ever an area of company operations was ripe for cost savings, here is one, and logistics points the way.

PURCHASING FUNCTION

Purchasing (or acquisition as this function is sometimes called) in this modern world has become far more than simply finding places to buy raw materials or subassemblies that can be worked up into finished goods. Let's use the following as our working definition of purchasing:

> PURCHASING: That aspect of materials management concerned with the movement of materials into and within a company.

In maintaining a sense of perspective, recall that materials management, along with physical distribution management, comprises the overall function we call "logistics." Purchasing is the part of the materials management branch of logistics whose objective is obtaining and moving materials at the lowest cost. Figure 2-1 provides an overview of the flow of materials and where the various company logistical functions fit into this stream.

In the "good old days" prior to World War II, purchasing had a much simpler job. Materials were both plentiful and cheap. Much has changed since then. Many types of materials are both expensive and, at times, difficult to get in the required quantities. Now, if the costs of purchased items are to be controlled, then some planning is necessary. Since planning is based on information, here is what is needed: The master production schedule tells the planners what the company is

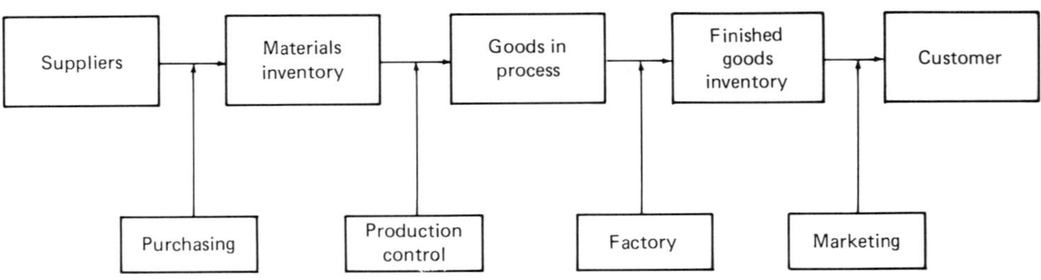

FIGURE 2-1
Flow of materials and logistical functions

going to make, the bill of materials is a rich source of information on what is needed to make it, inventory records tell the planners what the company has on hand to make the planned output, and materials requirements tell the company what inputs it must get.

Armed with the quantities of parts and components still needed to meet the production plan, purchasing can turn to locating sources of supply. This can be done through personal contacts between suppliers and company purchasing agents or advertisements in trade and professional journals. Finding vendors should include some estimate of the ability of the supplier to deliver items of acceptable quality at the time needed.

Why should the establishment of buying policies need to be mentioned? The necessity is not difficult to document. A great deal of disruption and just general frantic rushing around trying to get materials in (called "expediting" in days of yore) and orders out can be avoided if the company has a clear set of purchasing policies: what it is trying to do and the criteria of excellence in achieving the established goals (e.g., item and transportation costs, time standards for delivery, and quality-control standards).

Closely allied to the purchasing function is traffic and transportation. In the days before logistics became so important, transportation meant simply calling up a carrier for routes and rates. Now, even the proper description of items has a marked influence on ensuring lower freight rates. To give you an idea of how important the correct descriptions of items are on purchase orders, bills of lading, and allied shipping documents, consider that differences in the description of an item can affect freight costs by as much as 80 percent. Obtaining the lowest correct freight charge in view of the complexity of modern rate tariffs is no longer the function of an ill-informed clerk, especially since the routing of the shipment from the vendor to the plant rivals the selection of the proper freight rate in holding down transportation costs.

Despite some internal squabbles with other elements of the organization, persons concerned with inventory control have their own technical problems. A neat balancing act, not always done successfully, involves maintaining the lowest level of inventory consistent with efficient operation. An inspection function must be performed at the point of receipt in the company if quantity and quality are to be maintained. And if manufacturing is to be placated, inventory control should take responsibility for material flows within the organization to ensure the availability of materials, parts, and subassemblies when needed or, as is frequently the case, at a reasonable time delay.

Purchasing will need to coordinate closely with engineering on the problem of protective packaging. Marketing can attend to the type of packaging that attracts customers. If a decorative, attractive package will also protect the items from loss or damage in transit, fine! Artistically designed packages, however, may not always be strong enough to protect the contents; hence, an exterior pack of sturdy build, even if of pedestrian appearance, will be needed to keep the contents intact and meet the packaging requirements of transportation carriers.

Warehousing and materials handling are also functions closely allied to purchasing; they will be treated at greater length in Chapters 6 and 7. Nor should we ignore order processing, during which requests from the using elements in the organization are assembled. These elements could be the factory asking for materials to be worked on or inventory control flagging a reorder point. The purchase order is then prepared (with, remember, the proper description of the item ordered to ensure the lowest freight costs) and transmitted to the selected vendor.

PLACEMENT WITHIN THE ORGANIZATION

Purchasing, in common with other organizational elements, has two aspects: an informal and a formal placement. The informal organization is a loose but highly efficient information network consisting of personal alliances never seen on a formal organizational chart. People from purchasing may serve on a PTA committee with others from sales, on a bowling league with marketing representatives, and root for the Little League team with parents from manufacturing. Perhaps the single most characteristic feature of the informal organization in any company is that it does not follow the official organization chart. Often termed the "grapevine" (so called after the strings of military telegraph wires hung on trees along the roads during the Civil War), this informal network can carry more than just rumors. On occasions, it can be used to test the reaction to company policies without making public announcements that might have to be retracted. On other occasions, purchasing might get word over the grapevine of an unusual requirement developing in the factory long before the same information is transmitted from manufacturing via interoffice memo.

The formal organizational structure can take one of several forms depending on the size of the company. A small company may use a simple functional chart, as pictured in Figure 2-2. All major organizational elements are directly concerned with the product; thus the various managers report directly to the chief executive officer (the CEO).

A larger organization would divide its functional elements into two broad categories: line units, which engage in activities directly concerned with the product or service for which the company exists, and

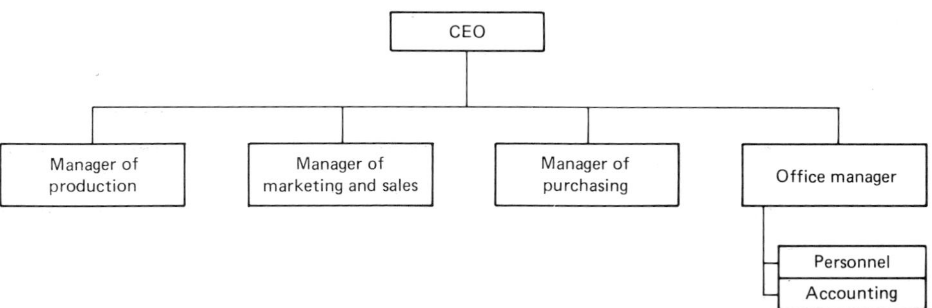

FIGURE 2-2
Organizational chart for a small company

staff units, which give advice, counsel, and assistance (but not direct orders) to the line elements, but do not work directly with the product. Some managers will perform a dual function as both line and staff managers. For example, the individual in charge of personnel in Figure 2-3

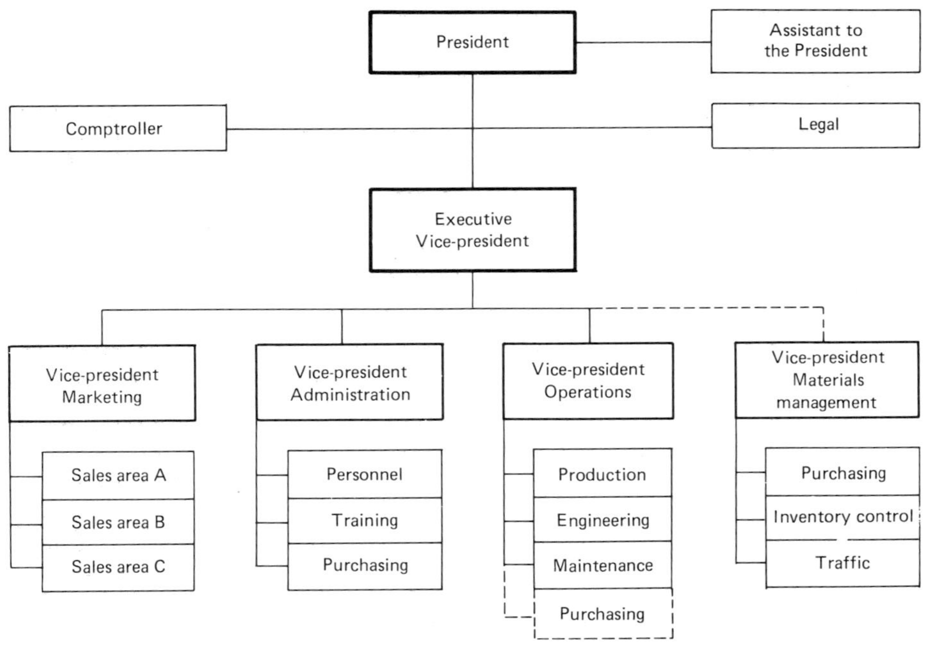

FIGURE 2-3
Organizational chart for a large company

performs a staff function for the rest of the company, but a line function within the personnel department.

In Figure 2-3, certainly operations would be considered line, and perhaps marketing as well. Administration, the assistant to the president, comptroller, and legal are just as clearly staff. A separate materials management element could be either line or staff; but since it occupies a position in the organization directly concerned with the physical handling of the product, classification as a line element would be appropriate. In fact, a better method is to organize materials management as a separate entity, as in the "Alternate placement" section of Figure 2-3, to emphasize the importance of this function.

Purchasing could be included as an element in either administration or operations, as indicated in the dotted portion of Figure 2-3, but a recommended placement would be under a separate vice-president for materials management.

Within the purchasing department itself, the various functions might be arranged as shown on Figure 2-4. Each company, of course, is going to organize itself to suit the nature of the business in which engaged. These organizational charts are merely suggested here in the full knowledge that an infinite number of variations are in use.

MATERIALS REQUIREMENTS PLANNING

Not all contact between purchasing and other elements of the company need involve conflicts of interest. One of the most fruitful areas in which costs can be conserved is in the planning that can be done with the production elements.

The demand for an item of inventory that purchasing is asked to acquire can be one of two types: independent demand for an item that does not depend on requirements for any other items, or dependent demand for an item whose requirements are determined by the end item in which it is used. Materials requirements planning (MRP) is used for items with dependent demand. As a result, the quantities of items ordered using MRP are based on projections of the number of items to be produced. MRP, then, can be defined in this way:

> MATERIALS REQUIREMENTS PLANNING (MRP): A scheduling technique for items whose demand is derived from (i.e., based on) the demand for a final product in which the item is used.

Items whose demand is original rather than derived can be planned using the economic order quantity (EOQ) techniques described in Chapter 9.

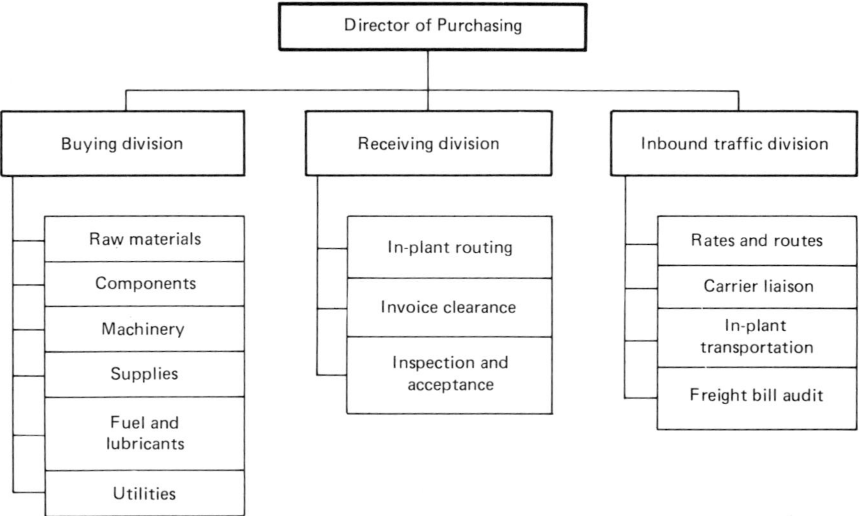

FIGURE 2-4
Organizational chart for a purchasing department

Since the demand for parts is based on the demand for the final end item, a bill of materials can be used to determine the quantities of parts needed and when they are used in the assembly operation. Armed with a forecast of the output of the final component, the quantities of parts required can be calculated. If the precise time at which the parts stock will be needed cannot be determined readily, then the *order-point* method of replenishing depleted parts can be used. This method involves placing a reorder when stock is drawn down to a specific quantity. If the production schedule is established well in advance, a different technique can be used, called *requirements planning.*

Essentially, requirements planning is a process of using production schedules to order in the quantities of parts that will maintain both a safety stock (to keep from running out completely if vendor deliveries are slow) and the quantities needed to keep production lines rolling. The best way to see how all this works is to trace an example through.

Materials requirements planning began in the 1950s and 1960s (like the Industrial Revolution, the exact beginning date is shrouded in the dim mists of history). As logisticians began to discover more uses to which the computer could be applied, MRP evolved into a scheduling technique rather than just a better way of ordering inventory. The expanded system required a good, very complete information system, which the computer was able to provide. When the cooperation of other elements in the organization was obtained, a rather comprehensive way of planning production became possible.

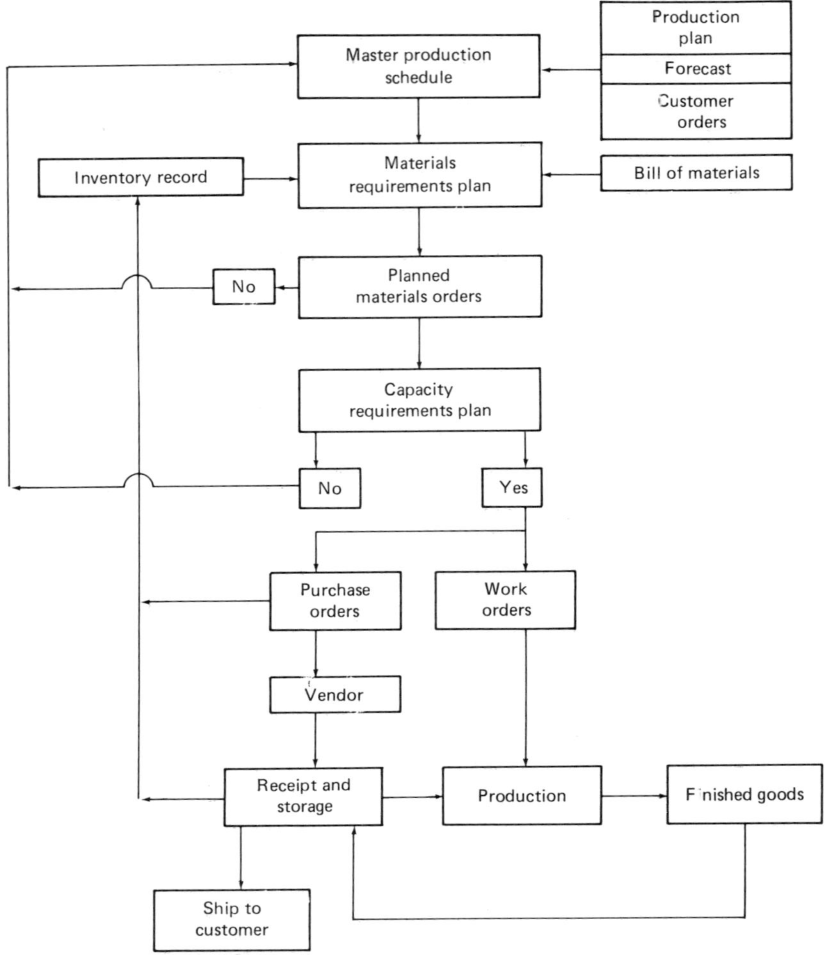

FIGURE 2-5
Closed-loop MRP

When the materials requirements plan is linked with the production plan and these two are matched against the capacity of the factory to produce the planned output, a variation on the basic MRP emerges known as *closed-loop MRP.* Figure 2-5 illustrates the components and the sequence of events in closed-loop MRP. The master production schedule, constructed from forecasts and customer orders, tells the logistician what is to be made. Inventory records inform the planners what is currently available, while the materials requirements plan tells everyone what must still be obtained to meet the production schedule.

Finally, the capacity plan tells the planners whether or not the plan can be accomplished. Other elements of the plan—issue of work orders to the factory and purchase orders to vendors, production of the finished products, and ultimate shipment to customers—help close the loop.

The loop nature of the plan is illustrated by incorporation of the capacity requirements plan. If labor, equipment, and raw materials are not available in sufficient quantities, the plan cannot be accomplished. The planning phase is a good time to discover these defects rather than during the manufacturing process. If, as should be, finance, marketing, data systems, and engineering are brought into the planning process, the resulting company-wide logistics systems is then known as MRP II.

At this point, let's leave the alphabet soup on the back burner and see how the materials requirements planning system works. Suppose you had the task of planning production requirements for a carburetor used in a large engine assembly. The following information is available to you:

- Forecasts of engine production each week are needed to make projections of the demand for carburetors. (Recognize here, the demand for carburetors is derived from the demand for engines.)
- A safety stock of 800 units should be maintained on hand to keep from stocking out should deliveries be slow or demand heavier than anticipated.
- The supplier of the carburetors requires a minimum order of 1,000 units.
- Required lead time to place and receive orders is two weeks.

Demand projections are entered in the production worksheet illustrated in Figure 2-6a. The production worksheet in Figure 2-6b shows how the worksheet is used.

Week 1: 1,800 units are on hand at the beginning of the week, but no deliveries are scheduled. The demand for 500 units draws the quantity on hand down to 1,300. Since the projected demand for 650 in week 2 would reduce stocks below the safety level, a purchase order for 1,000 carburetors is issued.

Week 2: As expected, this week's demand reduces carburetor stocks below the safety level of 800, but a shipment of 1,000 is due in next week to meet that problem.

Week 3: 650 units are on hand at the beginning of the week, but the 1,000 units ordered brings the total on hand to 1,650. The demand for 600 carburetors draws the stock down to 1,050. The projected demands for weeks 4 and 5, however, threaten to reduce stocks well below the safety level, so a purchase order for 2,000 units is issued.

PRODUCTION WORKSHEET

Item	Week					
	1	2	3	4	5	6
Quantity on hand at beginning of week	1,800					
Scheduled receipts	–					
Subtotal	1,800					
Projected demand	500	650	600	800	975	875
Quantity on hand at end of week						
Purchase orders issued						

(a)

PRODUCTION WORKSHEET

Item	Week					
	1	2	3	4	5	6
Quantity on hand at beginning of week	1,800	1,300	650	1,050	250	1,275
Scheduled receipts	–	–	1,000	–	2,000	–
Subtotal	1,800	1,300	1,650	1,050	2,250	1,275
Projected demand	500	650	600	800	975	875
Quantity on hand at end of week	1,300	650	1,050	250	1,275	400
Purchase orders issued	1,000		2,000		1,000	

(b)

FIGURE 2-6
(a) Production worksheet; (b) using a production worksheet

Week 4: As expected, the stocks by the end of this week have been reduced to 250 carburetors. This situation should be relieved with the arrival next week of the 2,000 units ordered in week 3.

Week 5: The 2,000 carburetors previously ordered brings the total stock up to 2,250 units, but the demand for this week reduces this total to 1,275. Since the demand projected for week 6 once again will reduce stocks below the safety level, an order for 1,000 more units is issued.

Week 6: Demand in this week will reduce carburetor stocks below the safety level, but the 1,000 new units ordered in week 5 will arrive in the next week to keep the production line going.

From the foregoing example, the continuous nature of requirements planning is obvious. Stocks of materials will not arrive by themselves, and thus require monitoring. This monitoring further requires close cooperation between the production and purchasing departments.

The Japanese use an interesting variation on this theme called *Kanban*. The Kanban technique seeks to operate with a minimum of inventory and use of space. In common with any system of inventory management, Kanban tries to keep inventories at the lowest level commensurate with good management. This method, however, keeps inventories at a minimum by producing only with what is available. In other words, in Kanban, inventory is a controlling factor in determining output. In MRP the reverse is true; production determines inventory requirements. More will be said about Kanban later as an example of change in logistics.

Anyone who has followed the preceding example in materials requirements planning might have a suggestion for both purchasing and production managers. If everyone is so worried about stocking out, why not buy a trainload of parts (and perhaps get a quantity discount as well) and thus have enough on hand to last for a year at a time. That's the trouble: having the big batch on hand for too long. Even a household that manages to consume several jars of the super crunchy peanut butter every week or so is not advised to haul a case home from the supermarket at one time.

First, a lot of space is taken up by the case, which could be devoted to stocks of other goodies like chocolate chip cookies. Second is the sheer cost of maintaining a large inventory of anything, whether it be chips, dips, or carburetors. The money tied up in the case of peanut butter is not available for other important uses, like making the mortgage payment, and thus the cost becomes prohibitively expensive. The same is true of a business. Inventory is expensive to buy, transport, and store. In consequence, people in Purchasing will never know any rest.

PURCHASING SEQUENCE

The process begins with a purchase order requisition submitted by some using element within the company. The requisition is usually triggered by the depletion of the stock of an item currently in use. This request could be received from production for a specific item or in the form of a bill of materials for an entire finished assembly about to enter a production run. Alternately, the requisition could be received from a warehouse or inventory control point as stock reached a reorder point. Whatever the source, information must now be gathered from which the purchase order itself will be prepared.

The *purchase order* is a formal order to a supplier to provide the goods in the quantities specified at the indicated time. This, of course, presupposes that a source of supply has been selected by competitive bid, by direct negotiation, or from a catalog, giving due consideration to price, availability, and inbound transportation costs. Once the supplier is known, the purchase order can be submitted based on descriptive information contained in the catalog or requisition. After the order has been dispatched, follow-up may be needed to determine how far along the supplier is in filling and shipping the order, or even expediting if the supplier is experiencing delays. Direct communication between vendor and buyer can be accelerated via teletypewriter or computer.

The ordering process, as may be expected, is not quite as simple as the brief foregoing outline. As an example of some of the complicating details, requisitions come in two general types: a *traveling requisition,* which contains a preprinted description of items usually ordered repeatedly, and the *descriptive requisition,* which is basically a blank form on which all information must be entered.

In an attempt to simplify the purchasing process, a number of specialized types of purchase ordering procedures are in use. The *blanket order* specifies a quantity of items that is to be delivered over a period of time, rather than all at once in a single shipment. Firms with several plants could consolidate all the purchases for the different plants with a single order called a *national contract.* In an effort to reduce the cost of carrying inventory, some companies use the *stockless purchasing system* for items purchased at frequent but irregular intervals. In this system, the items in inventory remain the property of the supplier until the stock is actually used. In *system contracting,* a long-term supply contract is negotiated that includes storage of the items purchased, thus saving on warehousing costs.

A rich area for cost savings is available in the procedures covering receipt of ordered goods. Fully 10 percent of all inbound shipments arrive either with incorrect documentation or no documentation at all. Imagine the confusion when a shipment arrives that no one can identify! Much of this is caused either by oral orders to the supplier, which have been misunderstood, or by simple clerical errors. In any event, unidentified shipments cause a confusion that is quite unnecessary.

To prevent this from happening, the receiving area should have on file and readily available a copy of a written confirmation of the purchase order from the supplier. This confirmation, which could be a copy of the purchase order with a signed acknowledgment by the supplier, should be on hand in the receiving area to assist in identifying shipments. Penalties should be imposed on suppliers who consistently ship late (perhaps by giving the next order to some other vendor), and the same should be true for suppliers who consistently ship early and thus transfer their inventory holding costs to their customers. These early shipments cannot simply be left on board the transportation equipment until a convenient time to unload can be found. Carriers charge a special penalty fee called *detention* in highway transportation and *demurrage* in rail transportation if the equipment is not unloaded within a specified number of hours.

In the receiving area, a clear distinction should be made between receiving a shipment and accepting it. The two words, receiving and accepting, are not synonymous in this connection. The receiving process should include identification by the copy of the confirmed purchase order, followed by inspection. Inspection of the shipment might include answers to such questions as these:

- Does a sample contain too many substandard items that have to be rejected?
- Are the items of such poor quality that they are not fit for "normal" use (i.e., cannot be used in items of the usual quality without reworking or, if used, result in end items of lesser than usual quality)?
- Do the items deviate materially from the specifications in the purchase order?

A yes answer to any of these questions by the inspectors should cause the shipment to be rejected and returned.

Rejection and return of unsatisfactory materials create an uncomfortable situation. The supplier must, by law, be given a chance to correct the defects, although some suppliers may balk at the rejection and claim defects in the standards prescribed in the purchase order. To avoid conflicts of this kind, the supplier should be required to sign an acknowledgment of the purchase order, thus creating a binding contract on the terms of the buyer. From this, you can see that the purchase order is not simply a blank form with a few spaces to be filled in.

The purchase order acquires many of the aspects of a legal document when prepared properly. The order should contain enough detail to make the terms on which items will be accepted unmistakable. These terms should include the technical detail needed to determine whether any deviations from the specifications are, in fact, material.

Everyone contributes to the formulation of "technical detail"—Production, Engineering, Logistics, and Legal. The nature of any desired warranties must be included and the time allowed for inspection specified. The purchase order is now becoming a rather detailed document, but necessarily so for the protection of the company. If the shipment must be rejected and returned to the vendor, remember to inform Accounting of the return so that the rejection may be confirmed with a *debit memo,* which tells the shipper that the amount owed is being reduced by the amount of the return.

PURCHASING IN PUBLIC AGENCIES

The purchasing function specifically, and logistics generally, in agencies supported by public funds is embedded in a legal environment far more confining than that of the private for-profit organization. For example, favoritism must never be shown in selecting suppliers (although frequent scandals show how imperfect human and legal systems can be), and equal access to the bidding process must be assured to all qualified bidders.

One of the strictest limitations on public purchasing is that imposed by the budget. To avoid exceeding the amounts of funds appropriated for specific purposes, purchase orders must be checked with unusual care and in almost every case must be certified by some finance or accounting office to confirm that the money is actually available.

The modern business environment is also amply supplied with legal restrictions that have a social context as well. Public agencies frequently are required by law to make maximum use of vendors whose owners belong to a minority grouping in society. Businesses not owned outright by a member of a minority group might still be required to submit proof that an adequate number of minority or disabled persons are employed, that the legal requirements concerning wages and hours are met, and that discrimination in hiring is not practiced before becoming vendors to public agencies.

Quite frequently, the purchasing function is highly specialized and centralized. The Treasury Department, for example, did the purchasing for the U.S. Government until the General Services Administration took over that function in 1949. The use of negotiated contracts rather than sealed bids is also quite common. The legal restrictions can be so numerous and complicated that the purchasing function in public agencies requires either the assistance of a lawyer or persons with extensive training in public procurement.

Whether lodged in nonprofit agencies or in for-profit private companies, purchasing is a central activity in the management of logistics. Unless engaged in that work, business logisticians do not necessarily have to be skilled purchasing managers, but they do need a detailed understanding of how logistics supports purchasing the better to perform their own jobs.

OBJECTIVES ACHIEVEMENT CHECKUP

The following exercises will help you in assessing the extent to which you have achieved the objectives for this chapter. Where essay-type answers are required, the more you are able to condense your knowledge into a short paragraph of a

few sentences the better. Looking back as you work these exercises and rereading parts of the chapter are permissible; but do not look ahead to the answers until you have done your best with the questions.

1. Match the function listed in Column II with the appropriate organizational element listed in Column I which performs that function.

Column I	*Column II*	
A. Purchasing	(a) Adding to the finished goods inventory	(a) ____
B. Manufacturing	(b) Locates sources of supply	(b) ____
C. Engineering	(c) Judges quality of purchased materials	(c) ____
D. Marketing	(d) Determines what is on hand to meet production needs	(d) ____
E. Finance and accounting	(e) Enforces quality standards on purchased materials	(e) ____
F. Inventory control	(f) Concerned with protective packaging	(f) ____
	(g) Order processing	(g) ____
	(h) Proper description of materials ordered from vendors	(h) ____
	(i) Designing a package to attract consumers	(i) ____
	(j) Certifying that items have been received from vendors	(j) ____

2. Why can't the purchasing function be performed in isolation more efficiently than if constant contact is maintained with organizational elements?
3. Differentiate between line elements and staff elements in structuring an organization.
4. Is purchasing a line or staff function?
5. List five of the various methods of ordering purchased materials.
6. How is logistics involved in ordering purchased materials?
7. You have been asked to use a production worksheet to help in ordering purchased materials and maintaining inventory on hand to prevent stockouts. The lead time in ordering from the vendor is 2 weeks and the minimum order the vendor will accept is 50 units. The safety stock has been set at 15 units. Use the worksheet in Figure 2-7a. Demand projections have already been entered on the worksheet as received from the factory.

PRODUCTION WORKSHEET

Item	Week					
	1	2	3	4	5	6
Quantity on hand at beginning of week	30					
Scheduled receipts	—					
Subtotal	30					
Projected demand	10	5	15	20	15	10
Quantity on hand at end of week						
Purchase orders issued						

(a)

FIGURE 2-7a
Practice production worksheet

PRODUCTION WORKSHEET

Item	Week					
	1	2	3	4	5	6
Quantity on hand at beginning of week	30	20	15	50	30	15
Scheduled receipts	—	—	50	—	—	50
Subtotal	30	20	65	50	30	65
Projected demand	10	5	15	20	15	10
Quantity on hand at end of week	20	15	50	30	15	55
Purchase orders issued	50			50		

(b)

FIGURE 2-7b
Answer to production worksheet problem

ANSWERS TO CHECKUP QUESTIONS

1. (a) B (f) C
   (b) A (g) A
   (c) B or C (h) A
   (d) F (i) D
   (e) A (j) A
2. A great deal of the purchasing function must be performed in cooperation with other elements of the organization if it is to be done effici-

ently and effectively. Operating in isolation would almost completely prevent the purchasing function from being performed at all. For some examples of this type of cooperation:

a. Transportation and traffic help in finding the lowest cost routing and carrier.

b. Production and inventory control provide data on what is in stock and due in.

c. Engineering works on protective packaging, which reduces loss and damage claims.

3. a. Line units are directly concerned (frequently actually "hands on") with production of the commodity for which the company exists.

   b. Staff units are not directly concerned with physical production but do give advice, assistance, and counsel to line personnel.

4. Technically, purchasing could be considered a staff agency since it is not directly concerned with production of the finished product. Purchasing, however, does have a line aspect since it is concerned with the physical handling of purchased input materials until turned over to the storage facility.

5. a. Traveling requisition

   b. Descriptive requisition

   c. Blanket order

   d. National contract

   e. Stockless purchasing

   f. System contracting

6. Logistics is involved in ordering purchased materials by being concerned with transportation, in-plant handling and storage, and recordkeeping.

7. Figure 2-7b offers a possible solution to the problem. Variations in assumptions concerning the reliability of the vendor to make deliveries on time and how much risk management is willing to take on stockouts would affect the solution.

3

Customer Service

CHAPTER OBJECTIVES

Upon completion of this chapter, you should be able to:

A. Understand the priorities between customer service and profits.
B. Differentiate between customer needs and wants.
C. Explain how standards of customer service can be determined.
D. Outline the major elements of the order cycle.
E. Appreciate the importance of distribution requirements planning.

If the primary purpose of any business organization is to meet the requirements of its customers, then customer service can be defined as:

> CUSTOMER SERVICE: All activities concerned with meeting customer needs and wants.

If you want to start a good argument, try debating about the primary purpose of business enterprise. Repeatedly, the chief executive officers of corporations acknowledge the subordination of profits to the requirements of the customer, for without buyers of the company's product, where are sales and the resulting profits? The logic is inescapable: The customer comes first. From the viewpoint of this chapter, service attracts and keeps customers just as much as quality of product. And without customers, no profit is possible. The need to tend to the needs and wants of customers is thus still paramount.

Two words have been used in the preceding paragraph: Needs and wants. The two are by no means synonymous. *Needs* refers to the basic customer requirements of low prices, acceptable quality, and service. *Service* would add to this getting the product to the customer when, where, and in the condition desired. This is commonly done through packaging, prompt order processing, and reliable delivery. Needs thus describes the minimum a firm must offer a customer in order to attract and hold business. Wants are in a different world.

Customer wants include a large variety of activities that add to the convenience and satisfaction of customers. Customer wants would include:

- Keeping customers informed of the status of their orders.
- Expediting emergency orders.
- Providing warranties and the repair or replacement promised by the warranty.
- Preparing workable contingency plans for filling orders during strikes, natural disasters, fires, or transportation delays and breakdowns.

Certainly, when an order is received, customer service should shift into high gear. Attention, however, to the needs and wants of customers can be carried both forward and backward, starting at the moment an order is received.

Even before an order is placed, the vendor should have a clear idea concerning the exact content of the customer service to be provided. This customer service policy should include definite objectives on delivery times, back-ordering procedures for items temporarily out of stock, the precise manner in which returns are to be handled, and the contingency plans that ensure the continuity of customer service in times of emergencies. Perhaps most important of all, the customer should be informed of this policy. The company is further obligated to ensure that the customers understand this policy.

When the order is received, inventory management procedures should be efficient enough to guarantee sufficient stock on hand to fill the order. Obviously, orders are to be filled accurately and the most effective mode of transportation selected. This may mean selecting the most cost-effective mode of transportation that balances speed against cost. Of course, if the customer is in a hurry and willing to pay, premium transportation can be used for especially quick delivery. If the exact item ordered is not on hand, the company should have a way of soliciting from the customer a willingness to accept substitutions of items or adjustments in the quantities to be shipped.

After the order has been shipped, a new set of customer service needs appears. Things do go wrong with shipments despite the best of care and handling. Complaints must be heeded and very specific procedures followed in authorizing and receiving returned goods.

Repair parts present a particularly difficult problem. First, an adequate supply of parts must be kept on hand to service models currently in production and being sold. But what about parts for models whose production was discontinued years ago? For how old a model should repair parts be maintained in stock? Parts for the old Ford Model T might be rather difficult to obtain today, yet a small demand still exists for those ancient cars still running. Considering that inventory represents money prevented from being put to other uses, the age of the model for which parts will be made available from current stockage is a sticky problem for management to solve. How good for the public image of the company is the ability to meet a demand for parts no matter what the age of the model! But how expensive and wasteful of storage space is the maintenance of this type of public relations! Conceivably, management might choose image over solvency, but that does not make the choice any less difficult.

Experimental models not in full production present the same problem. The degree of support for outdated or exotic equipment requires the same balance between cost and to what extent customer demands will be met. Bringing up the end of this list is the provision of facilities for prompt repair or replacement under warranties.

HOW MUCH CUSTOMER SERVICE TO HAVE

Airy generalizations about the necessity for meticulous attention to customer needs is not enough; the standards of customer service should be both clear and quite specific. Standards can be expressed in a number of ways according to the nature of the product:

- No more than three shipment errors per 1,000 shipments.
- Delivery of parts within 36 hours.
- Ninety-eight percent of customers' orders ready for shipment within 24 hours.
- Fourth morning delivery for all orders.
- No more than 5 percent stockouts per year.

These standards, when understood by employees and customers alike, should be enforced rigorously. For example, in all the burgers eaten at McDonald's, have you ever received a bun with a hole poked in it? Never. At McDonald's, a slip of the thumb means a discarded bun.

CUSTOMER SERVICE AND DIMINISHING RETURNS

A high degree of customer service is a noble aspiration; but as service improves, costs eventually exceed gains. In other words, customer service is subject to diminishing returns (a principle well known to students of economics):

> PRINCIPLE OF DIMINISHING RETURNS (as applied to customer service): As resources are expended to improve customer service, sharp increases in sales or reductions in customer complaints are experienced initially; but as resources continue to be applied, gains begin to diminish and eventually cease.

Diminishing returns appear in three stages: increasing returns, diminishing returns, and negative returns.

On the graph pictured in Figure 3-1, the stage of increasing returns appears at the far left. The customer service curve has an increasingly steep slope, which means the investment of resources (e.g., more money or employees) in improving customer service is paying off handsomely with noticeable increases in sales, fewer merchandise returns, less complaints, or whatever criterion of excellence is being used. We might say that the increase in gains is proportionally greater than the increase in resources used to achieve the gains.

Next, as more resources are expended, the rate of improvement, as evidenced by the slope of the service curve, gets lesser. At the point on the service curve at which the rate of increase becomes steady, the stage of diminishing returns has been entered. This indicates that the criterion of excellence is showing a gain that is proportionally less than the increase in the use of resources. Perhaps the rate of increase of sales is slowing at this point, or the rate of decrease in customer complaints is lessening.

Finally, as still more effort is expended in improving customer service, the stage of negative returns is entered. At this point, where the

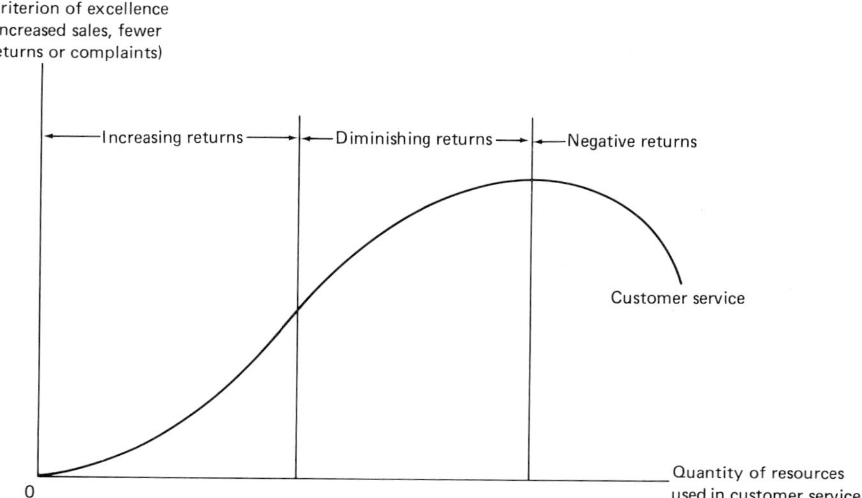

FIGURE 3-1
Diminishing returns

service curve turns downward, too much effort is being expended. Service calls to check up on customer satisfaction are becoming burdensome to the customer as salespeople make pests of themselves; too much inventory is being maintained to meet customer demands, too many storage locations have been established to speed deliveries, too much premium transportation is draining off profits. In general, rather than experiencing gains in customer satisfaction, the additional effort is causing customer dissatisfaction. Logistics managers thus need to know when the area of diminishing returns is being entered—or when the company has enough customer service to meet the demands of the customers in a timely fashion and minimize stockouts, transportation delays, and other customer irritants.

ORDER CYCLE

A major determinant of customer service is the order cycle. The main elements of the order cycle are as follows: placement, processing, and delivery. The flow of the cycle is illustrated in Figure 3-2.

Orders can be placed by customers using the telephone or mail or by delivery of a purchase order. Electronic transmission of orders has become possible as computers talk to computers, with human monitors using display screens checking on the information passing between the machines.

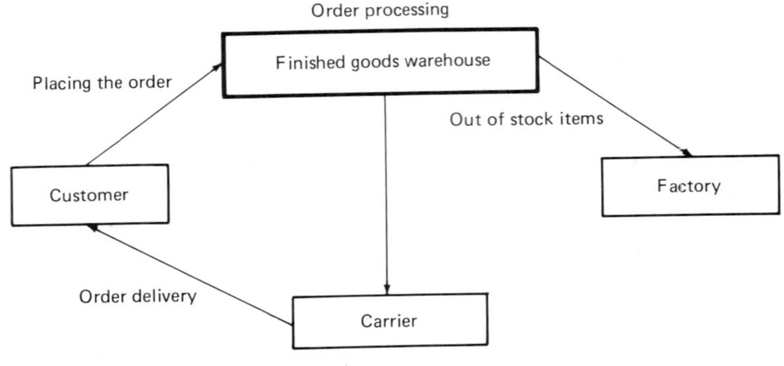

FIGURE 3-2
Order cycle

Processing the order after receipt requires several separate activities. While the order is being checked for the availability of items, the finance department will be running a credit check on the customer. If the customer has a good payment record (or good credit rating if a new client), accounting can record the transaction. This check on credit worthiness should not be minimized. Many business persons tend to ignore that much commercial activity is not conducted with funds obtained by bank loans. Instead, companies lend to each other in the form of trade credit. Merchandise is bought "on account" (or, as the accountants term it, by the opening of an account receivable on the books of the seller and an account payable on the books of the buyer), with cash payments made later. Thus a credit check is imperative to prevent the vendor from being burdened with a large number of what accountants delicately call "doubtful accounts."

Once the order has been released by the finance department, accounting makes the appropriate entries on the books of the selling company and the logistics elements begin filling the order. Inventory control instructs the warehouse or branch warehouse (often called a distribution center) that has the needed item in stock and is nearest to the customer to pick and pack the order. If the item is not in stock, a production order is issued to the factory.

When the warehouse or distribution center reports the order ready for shipment, a carrier is selected, thus involving the transportation and traffic department. The transportation equipment is loaded, and the rail car, truck, barge, or plane is released to the carrier. The more efficiently the order cycle is managed, the shorter the elapsed time and the happier the customer.

Ideally, customer service should contribute the maximum possible to profit; but the point of maximum profit contribution may not be

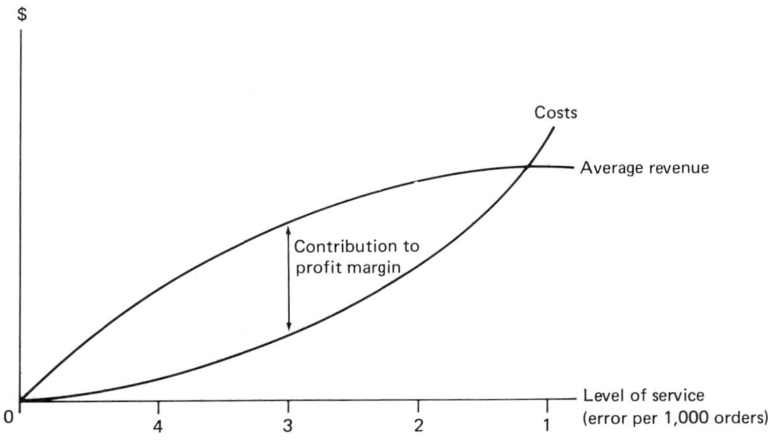

FIGURE 3-3
Cost of customer service

easy to find. Customer service really represents a trade-off between cost and profit. Suppose the level of customer service was to be measured by the number of errors per 1,000 orders. Reducing errors involves a cost to the company, which tends to rise as the number of errors is reduced. Initial improvements can be dramatic on occasions. When the Matsushita Electric Company bought out the Motorola television plant in Illinois, a rate of 150 defects for every 100 sets produced was being experienced. The new managers brought this defect rate down to 3 or 4 per 100 sets. As more costs are expended improving customer service, a point will be reached at which further improvement is simply not worth the cost involved. This is illustrated in Figure 3-3.

As additional resources bring the error rate down, the contribution to profit initially increases. As the error rate continues to fall, the costs involved rise and eventually catch up with revenue until no contribution at all is made to profit. In other words, a 100 percent error-free operation would be nice to have, but it is not cost effective. On the graph in Figure 3-3, an error rate of about 3 per 1,000 makes the greatest contribution to profit. In this example, error rates less than 3 per 1,000 become so expensive to achieve that the profit contribution narrows markedly.

HANDLING RETURNS

Customers return items for a large variety of reasons: the wrong quantity was shipped, damage in transit, the wrong part was ordered, a customer of the buyer returned the item, or a product-recall program required the return. The last reason, a product-recall program, opens up a whole new world of problems. First, the ownership of the product must be traced. This can be done by letting the computer trace the origin of each order through the warranty cards on file. Then the item must be retrieved. Finally, the defective item is either repaired or replaced and returned to the customer. The massive recalls that occasionally occur in the automotive industry are evidence of the difficulty of conducting such a campaign.

Once the item has been received, management must decide what to do with it. If the returned item is not part of a recall program, the returned goods need to be inspected. In the event the customer erroneously returns a perfectly good part, the shipment should be refused. If the returned item is accepted, should it be reworked and either returned to the customer or stock, should it be scrapped, or is it possible to cannibalize the usable parts from an otherwise defective assembly? Management undoubtedly will want to get the engineers involved in handling returned goods in addition to the regular logistical staff.

MEASURING CUSTOMER SERVICE

Customers should be served in the most expeditious manner possible; but each customer has a different set of needs. Customers who buy raw materials or semifinished goods for further processing and assembly generally prefer reliability in delivery to speed. Production scheduling techniques can allow for speed of delivery in specifying the lead time; but a delivery schedule met without exceptions month after month would be a great asset to any manufacturing operation. Knowing that a vendor can be counted on to deliver as promised, even though the speed of filling the order might be a bit slow, is a highly desirable attribute among suppliers.

Customers in the retail trade usually prefer availability at the times needed. The inventory-control procedures of such customers can compensate for slow speed of delivery and minor variations in time of delivery. Being able to obtain the item when orders are pouring in is much more to be desired than sheer speed of delivery.

In further contrast, customers who deal in products with short shelf lives, such as restaurants offering highly perishable dishes, medical facilities using radioactive isotopes with short half-lives, or drug products with expiration dates, much prefer speed of delivery and reliability. Yet a company cannot construct a single standard of service for each of perhaps several hundred individual product lines. Thus, in many cases, the discussion of customer service boils down to the fundamental question, How long does it take to get an order filled and delivered?

The shorter the time involved, the less cost required to carry the inventory and the happier the customer becomes. Just how short this time should be presents a problem in assessing the proper level of customer service to be offered. Very short order cycles require a high level of inventory, a larger work force to pick and pack the order, and perhaps the use of premium transportation if the order is shipped late. How high the level of customer service should be depends not only on what the seller is able to provide, but also on what the customer wants and is willing to pay for.

DISTRIBUTION REQUIREMENTS PLANNING

In Chapter 8 we deal at greater length with the problem of forecasting and some of the techniques used. Once a forecast of demand has been made, how can it be used to improve and sustain customer service? One answer is the use of distribution requirements planning (DRP).

> DISTRIBUTION REQUIREMENTS PLANNING: A technique for matching inventory availability with customer needs.

DRP is concerned with matching delivery schedules and customers needs with finished goods inventories. This should be contrasted with materials requirements planning, which is concerned with inventories of goods used in the process of manufacturing.

In using DRP, each branch warehouse or distribution center submits its forecast of customer needs (both anticipated from past experience and from customer orders actually on file) to a central location, perhaps a main warehouse or storage facility. When the consolidated forecasts are furnished to the factory, a production schedule can be established using a master production schedule. This technique aids in the planning process, since managers can see immediately from the forecasts where adjustments might be required in the production and delivery schedules. Where inventory is being controlled by the specification of a safety level rather than a production schedule (i.e., a fixed reorder point system), the use of DRP permits control of the reorders and the arrival schedule. In addition, DRP can also handle unexpected variations in customer demand.

As an example of using DRP, suppose the main warehouse specified orders from distribution centers to be in lots of 100 units. Or the factory might prefer work orders in multiples of 100 units. A safety level of stock is set at 50 units, and a lead time of two weeks is considered about normal for arrival of reordered stock. The distribution requirements planning worksheet might look like the format illustrated in Figure 3-4a.

DISTRIBUTION REQUIREMENTS PLANNING WORKSHEET

Item	Week						
	1	2	3	4	5	6	7
Demand forecast	20	50	30	40	50	20	35
On hand at beginning of period							
Receipts							
Total on hand							
Customer orders		(40)			(100)	(120)	
On hand at end of period							
Amount ordered							

(a)

DISTRIBUTION REQUIREMENTS PLANNING WORKSHEET

Item	Week						
	1	2	3	4	5	6	7
Demand forecast	20	50	30	40	50	20	35
On hand at beginning of period	50	120	80	30	130	30	10
Receipts	100	–	–	100	–	100	100
Total on hand	150	120	80	130	130	130	110
Customer orders	30	(40)	50	–	(100)	(120)	35
On hand at end of period	120	80	30	130	30	10	75
Amount ordered	–	100	–	100	100	–	100

(b)

FIGURE 3-4
(a) DRP worksheet; (b) using a DRP worksheet

Initially, the demand forecasts of the distribution centers have been totaled and entered on the first line of the worksheet for each week. Customer orders actually received have also been entered on the worksheet on the "Customer orders" line and circled to distinguish them from orders received during the week on an unplanned basis. The sequence of events for the next few weeks would be entered on Figure 3-4b.

Week 1: An unexpected customer order for 30 units can be met easily from the existing stock as can the reorder due in this week. The customer order for 40 units already on hand for week 2 also is no problem.

Week 2: Actual customer orders for 40 units this week and the 100 ordered for delivery in week 5, plus any unplanned orders that might arrive, could present a problem. As a result, a replenishment order for 100 units is placed this week.

Week 3: As might be expected, an unplanned customer order for 50 units arrives this week, compelling a dip into the safety stock. But next week, no orders are on hand and the replenishment stock ordered in week 2 should arrive.

Week 4: A slow week. The replenishment order arrives, putting the inventory of finished goods in good condition for the known customer orders of 100 units that must be filled next week. The known demand for 120 items in week 6 might be troublesome, so another replenishment order for 100 units is placed to arrive in week 6.

Week 5: The known customer order for 100 units due out this week is shipped, causing a penetration of the safety stock. The replenishment order of 100 units will arrive next week and enable us to handle the known customer order for week 6. Activity in week 6, however, could reduce stocks below the safety level, so another replenishment order is placed.

Week 6: The known customer orders for 120 units are shipped out of the 30 units on hand and the newly arrived 100 units ordered in week 4. Stocks are now very low, but the replenishment order placed in week 5 for 100 units and due to arrive next week should prevent a stockout.

Week 7: The replenishment order arrives on schedule, but so does an unplanned customer order for 35 units. By the end of the week, the safety stocks still have not been tapped; but placing a replenishment order now will help handle customer orders beyond this week.

Notice in this example that the forecasts received from the distribution centers were not especially accurate; but this is not unusual in a changing business climate. The use of DRP aids considerably in planning production to meet these recorded and unanticipated demands. Changes in the forecasts and actual demands can be reflected in changes in the materials requirements planning system, which can then be linked to both production schedules and the purchasing of raw materials for production.

One bonus in using distribution requirements planning is the information it provides to revise forecasts if the existing demand projections prove to be wildly inaccurate. Such a revision in forecasts, or even a change in forecasting techniques should that be required, is vitally necessary because the DRP system could break down if the gap between sales forecasts and actual customer orders is too wide over too long a period of time.

With our own channels of supply and distribution in reasonably good order, we are now free to address the problem of transportation, or determining precisely how raw materials and finished goods are carried about the landscape to meet materials and distribution requirements.

OBJECTIVES ACHIEVEMENT CHECKUP

The following exercises will help you in assessing the extent to which you have achieved the objectives for this chapter. Where essay-type answers are required, the more you are able to condense your knowledge into a short paragraph of a few sentences the better. Looking back as you work these exercises and rereading parts of the chapter are permissible; but do not look ahead to the answers until you have done your best with the questions.

1. Should customer service have priority over profits? Explain.
2. Which is more basic, customer needs or customer wants?
3. Comment on this statement: "You can never have too much customer service."
4. Describe the order cycle.
5. Differentiate between materials requirements planning (MRP) and distribution requirements planning (DRP).

ANSWERS TO CHECKUP QUESTIONS

1. Yes, because unless the customer is satisfied and continues to buy the product, no profits will be earned.
2. Customer needs are more basic because these involve such essential elements as the price, quality, quantity, and service required to attract and hold customers. Wants involve somewhat less pressing matters such as convenience and comfort.
3. The principle of diminishing returns is applicable here. Devoting an increasing amount of resources to improving customer service eventually reaches a point at which more customer service produces proportionately less improvement in customer satisfaction.
4. Initially, the order is received by the vendor from the company ordering the products. The order is then processed and filled (with assistance from finance and accounting to check on the credit worthiness of the customer, inventory control, which starts the picking and packing process, and traffic when the order is ready to be loaded on a carrier) and finally shipped out.
5. MRP is concerned with making parts and subassemblies available when needed by production. This usually involves a close and complex interaction between production, inventory control, marketing and sales, transportation, engineering, and almost every department of the firm concerned with making and selling the product. DRP is concerned with ensuring the availability of stocks when needed by buyers of the finished product. DRP also requires a close linkage with other logistical elements of the firm.

4

The Transportation System

CHAPTER OBJECTIVES

Upon completion of this chapter, you should be able to:

A. Appreciate the importance of transportation to the national and world economy.
B. Describe the major characteristics of the modes of transportation.
C. List the transportation functions included in logistics.
D. Explain the transportation bridges.
E. Understand the various kinds of intermodal transportation services.

No one really seems to get excited about transportation until the availability of the system is threatened. A truckers' strike decimates the supply of broccoli and other vegetables in the supermarkets and suddenly backyard gardens blossom with salad ingredients all over the neighborhood. An airliner crashes with all hands on board, and vacation thoughts abound concerning the virtues of the zoo and other local entertainments for the family. Rail freight service deteriorates causing the thoughts of company traffic managers to turn to the purchase of company-owned fleets of trucks. Yet, for most of the time, the transportation system of the nation operates well on hours that would send a company union into instant revolt: twenty-four hours a day, seven days a week, with precious little time off for holidays. Transportation is one of the few industries that cannot shut down on Friday afternoon and resume business Monday morning.

WHERE DID IT ALL BEGIN?

The Colonial era was one of birchbark canoes and crude river barges. Coastal shipping and smaller vessels on the rivers and waterways predominated. What about overland transportation? Choose your mode: a slow, leisurely but smooth journey on a riverboat or a bone-jolting ride in a lightly sprung coach over the ruts that passed for roads and trails. As a result, very little freight moved overland.

Until Robert Fulton installed a steam plant in a river boat, the barge spelled freight transportation in early America. After invention of the steamboat in 1811, times changed. The steamboat and barge shared the burden of hauling passengers and freight on the rivers and lakes until the railroads could extend their lines and provide a faster and eventually cheaper alternative.

After the railroads began chugging the 14 miles from Baltimore to Ellicott City, the steam locomotive became the transportation king of American transportation. The king was dethroned by the diesel locomotive in the 1930s, which was then promptly unseated by highway and air transportation. The rails lost their passengers to the airlines and their freight traffic to the motor carriers. And what a pity! Lost to future generations is the sadness of the wail of the steam locomotive passing in the night and the thrill of the overland express headed by an engine belching impressive clouds of steam, smoke—and air pollution. But the real pity is that no other means of transportation can haul such large quantities of bulk freight as efficiently and economically as rail.

To round out this thumbnail developmental history: The gooey glop skimmed off the surface of some ponds in Pennsylvania and bottled as the panacea "snake oil" ultimately found its way into petroleum pipelines shortly after the close of the Civil War.

WHAT CAN TRANSPORTATION DO FOR YOU?

The cost of transportation uses about 20 cents out of every dollar spent in stores to cover the cost of hauling the product to you. The bite out of the logistical dollar contributed by business firms is even larger: Two-thirds of each logistical dollar gets spent on some form of transportation service. Transportation is thus something the logistics manager is not about to ignore. The availability of transportation can have a significant effect on plant and warehouse location, while the specific mode of transportation available directly affects inventory management. A slow mode of transportation (which also usually has the advantage of being cheap) will dictate large inventories to keep goods on hand between the slow arrivals. A rapid mode of transportation (usually also rather expensive) will dictate smaller inventories since resupply is quicker.

The packaging required is often prescribed in some detail in the tariff publications of the modes of transportation being used by the company. In consequence, these requirements become a matter for engineering. The type of equipment furnished by the carrier will also have an impact on the type of equipment the company will acquire to load and unload materials. The mode selected will influence the degree to which the corporate goals of customer service will be met, or at least give some indication concerning how difficult it will be to maintain the desired level of customer service. Attention paid to the structure of freight rates will be repaid with lower freight bills. For example, the shipment of goods in carload lots is much cheaper than the smaller less-than-carload shipments. In addition, using carload quantities could also mean obtaining quantity discounts from suppliers. All these considerations enter the deliberations of the logistics manager when contemplating the use of transportation.

Table 4-1 summarizes the rankings of the major modes of transportation for various characteristics. In interpreting the entries in Table 4-1, a 1 is highest and a 5 is lowest. Thus, air ranks first in cost (i.e., is the most costly mode of transportation), first in speed (since air is generally the fastest mode), third in dependability (i.e., is about average since flights can be delayed for any number of reasons besides skyjackers), fourth in capability (i.e., air does not have a capacity to carry just any kind of freight in the same sense that a greater variety of products can be loaded aboard a freight car or trailer), and fifth in fuel efficiency of all

TABLE 4-1
Modes of Transportation

Characteristic	*Rail*	*Air*	*Highway*	*Pipeline*	*Water*
Cost	2.9¢/ ton-mile	$45/ ton-mile	15.3¢/ ton-mile	Under 1¢/ ton-mile	0.45¢/ton-mile on rivers 0.52¢/ton-mile on Great Lakes
Cost ranking	3	1	2	4	5
Speed	3	1	2	5	4
Dependability	4	3	2	1	5
Capability	2	4	3	5	1
Fuel efficiency	3	5	4	1	2
Loss or damage	5	3	4	1	2

the modes of transportation (i.e., is the most fuel inefficient of the modes).

MODES OF TRANSPORTATION

Carriers vary considerably in their characteristics. Speed ranges from the supersonic in the air to a leisurely 5 miles per hour by river barge. Shipments can vary in size from 150 tons in a railroad boxcar to only 10 tons shipped by truck. A broad spectrum of equipment is available depending on the type of products carried. Prices also vary, as do the types of services offered.

The logistics manager thus has a lot to think about when contemplating selecting a mode of transportation for the products of the company. Included in the general area of logistics are such transportation functions as:

- Negotiating rates and routes.
- Selecting routes and carriers.
- Appearing before regulatory agencies to support or protest a change in rates affecting the company.
- Evaluating carrier performance.
- Analyzing transportation costs and services.
- Operating company-owned means of freight and passenger transportation.

- Filing loss and damage claims.
- Auditing freight bills to ensure that the proper charges were paid to carriers.

And don't assume that this list exhausts all possibilities. The situations in which individual companies find themselves are so enormously varied that a thorough study must be made of the transportation needs of each firm to determine which transportation functions apply.

Carriers can be grouped into four major categories:

- Common carriers, who hold out their services to all who wish to use them.
- Contract carriers, who haul freight for individual companies on a contract basis.
- Private carriers, who own both the equipment they use and the freight carried.
- Exempt carriers, who haul only specific types of products (e.g., unprocessed, raw agricultural products).

Common carriers tend to be the most closely regulated by public authorities, contract carriers less so, with private and exempt carriers enjoying the most freedom from economic regulation. All types, however, must observe the laws and regulations governing safety.

Railroads

Rail has pride of place in the history of transportation in almost every country of the world. Germany, for example, became a powerful nation-state only after the Prussian chancellor Bismarck stitched the separate Germanic principalities together with a system of railroads. The development of the United States was largely confined to the eastern seaboard until the railroads opened up the rest of the nation to settlement and commerce.

Railroads have been divided into three classifications according to the operating revenues earned each year:

- Class I railroads earn operating revenues over $50 million annually. Less than 40 are now in existence and operate 171,178 miles of track.
- Class II railroads earn operating revenues ranging between $10 and $50 million annually. Less than 30 are now in existence.
- Class III railroads earn operating revenues of $10 million or less annually. Railroads in this category are usually switching and terminal companies.

One of the major advantages of rail is the ability to haul large quantities of products over long distances. For example, unit trains of up to 100 cars haul coal from the mines in West Virginia to the coal-loading pier at Newport News, Virginia.

Speed on the rails averages about 20 miles per hour, but this speed is maintained night and day with stoppages only to load fuel and water, change crews, and pick up and discharge freight. A big advantage of rail is the ability to interchange cars without unloading and reloading the freight. For example, as a passenger you cannot travel coast to coast in the same rail car; but a hog can be loaded in a freight car on the East Coast and stay in the same car all the way to the West Coast. Somewhere—at Chicago, St. Louis, New Orleans, or some other gateway—you will have to change trains. But not the hog. The pig's freight car is simply switched from one rail line to another without inconveniencing the animal.

Several technological innovations in recent years have greatly improved rail service. Car shortages are a constant headache to both railroads and shippers. During the harvest season, for example, hopper cars to carry the grain from the growing areas to the grain elevators are in short supply. In 1974, Railbox was formed, which provides a nationwide fleet of over 25,000 general-service box cars. Since the cars belong to no single railroad company, they can be shunted around the country to the areas of greatest need.

Classification yards are now used to break down inbound trains and reform the cars into outbound trains according to destinations. This procedure had been a significant source of delay. Computers are now used in which optical sensors read coded labels on the sides of cars and thereby speed up the classification of cars. Run-through train service is now available in which the trains bypass the classification yards en route and pass directly from origin to destination. Unit trains carry large numbers of cars loaded with a single product, such as grain or coal. Nor should we ignore the welded rail, that single ribbon of steel without breaks or joints extending for miles that lessens the damage to freight due to vibration.

These innovations are especially helpful considering the type of product carried on most rail shipments. About one-fourth of all the freight hauled by rail is coal. Another 8 percent consists of agricultural products. Any important improvement in service eventually translates into dollars and cents saved through lower freight rates and less costs to shippers in better handling of their products.

The railroad industry, of course is not without its problems. When revenues run low, economies are obtained in operating costs by slighting the maintenance of way and equipment. The old Paoli Local, which served the Philadelphia suburbs for generations, was renowned for the shabby equipment in which the passengers were jammed due to lack of adequate maintenance. Labor stoppages originating in poor labor relations are not unknown (although some of the earliest unions were in the railroad industry). Still, rail reigns supreme as an economic and efficient hauler of bulky items over long distances.

Motor Carriers

Motor carriers are also divided into three categories according to the annual operating revenues earned:

- Class I motor carriers earn gross operating revenues over $5 million annually.
- Class II motor carriers earn gross operating revenues between $1 and $5 million annually.
- Class III motor carriers earn gross operating revenues of less than $1 million annually.

Another way of classifying motor carriers is by their legal status:

- Common carriers, who must serve all who ask their services (provided, of course, the carrier has the necessary equipment).
- Contract carriers, who haul freight for individual shippers under specific written agreements.
- Private carriers, who own the freight they haul.
- Exempt carriers, who haul farm products, fish, or livestock or who operate within the confines of a single city. Note that any carrier becomes exempt when hauling the named products.
- Brokers, who own and operate no equipment but bring together (as any broker does) those who wish to ship freight and those who wish to haul it.

Highway transportation can be extremely flexible. If a local carrier does not have the exact type of equipment needed to haul a particular commodity, some other carrier not far distant will have the necessary specialized equipment. This ability to tailor the service to the specific type of traffic means added convenience to shippers. Truckers offer both truckload and less-than-truckload service, but most shipments by highway seem to be of the small, less-than-truckload variety.

Motor carriers also have the advantage of not being required to build and maintain their own right of way. In fairness, although the public pays for the construction and maintenance of the highways, motor carriers pay out large sums each year in road-use taxes, which are then used to rebuild and maintain the roads. The highway system itself is quite extensive, to the advantage of both truck lines and shippers. The United States has about 3.8 million miles of roads, streets, and highways, while the interstate system accounts for 42,460 miles of this. The system of interstate highways, however, carries 20 percent of all motor freight traffic.

Quality of service is a constant problem to motor carriers. Loss and damage claims tend to be high, and frequently service is slow due to the necessity of rehandling large volumes of small shipments at transfer points. Pickup and delivery service is often unsatisfactory from the viewpoint of the shippers. All this adds up to a loss of traffic by common carriers to private carriers. And still the common carriers ask for higher rates.

Some Economic Aspects of Rates

Let's pause here for a bit of economic philosophy. Since shippers are permitted to attend rate hearings and submit their contributions to the discussion, some background might provide a better basis for such intervention.

Will higher rates always generate the increased income carriers claim they need? The answer is: It all depends. This sounds perilously close to no answer at all. It all depends on what? It all depends on the elasticity of demand for the particular transportation service.

> ELASTICITY OF DEMAND measures the responsiveness of quantity of service sold to changes in rates.

Demand is *inelastic* when a change in rates (either up or down) causes a less than proportional change in the volume of traffic. Demand for a particular transportation service would be inelastic if a 5 percent change in rate caused a 2 percent change in the volume of traffic carried. At the other extreme, demand is *elastic* when a change in rates (again either up or down) causes a greater than proportional change in the volume of traffic. Thus, demand for a particular transportation service could be elastic if a 5 percent change in rates caused a 9 percent change in the quantity of traffic carried. With these ideas in mind, look at the question just posed.

Suppose the demand for the service of transportation is highly inelastic, as portrayed in Figure 4-1. An increase in freight rates from r_1 to r_2 will cause a decline in the volume of traffic from q_1 back to the smaller q_2. Note that the change in traffic volume is considerably

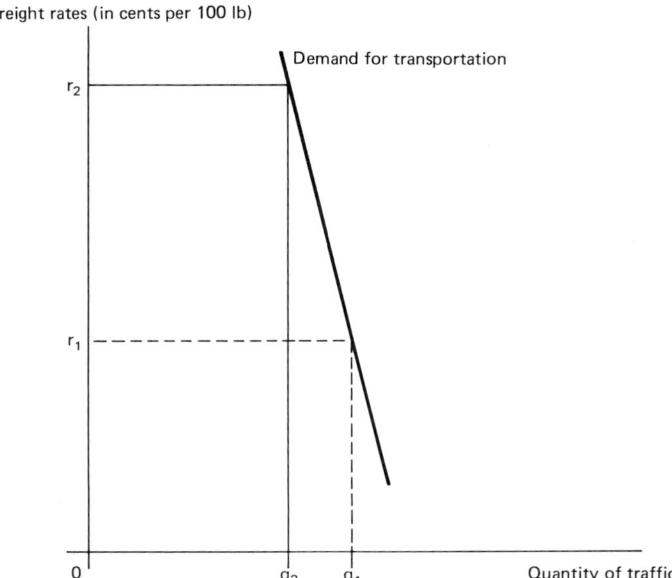

FIGURE 4-1
Freight revenue changes with inelastic demand

smaller than the change in rates. Certainly, the decline in the quantity of traffic carried will cause revenue to decrease; but the increase in rates is so large it overpowers the effect of the decreased traffic, and the end result is an *increase* in revenue.

Suppose now that the demand for the service of transportation is highly elastic, as portrayed in Figure 4-2. An increase in rates from r_1 to r_2 will cause a sharp decrease in the quantity of traffic from q_1 to q_2. Note that the change in traffic volume is considerably greater than proportional to the change in rates. Certainly, the increased rates will add to revenue, but the decline in the quantity of traffic carried is so large it overpowers the effect of the increased rates, and the end result is a *decrease* in revenue.

What affects the elasticity of demand for transportation service? The elasticity of demand for the product being hauled. In other words, if no demand exists for the product, no demand will exist for the transportation of the product. Therefore, if demand for the product is elastic, demand for the transportation of that product will also be elastic. To carry this analysis one step farther, what has the most influence on the elasticity of demand for both the product and the carriage of that product? Primarily, the availability of substitutes. If the product is bubble

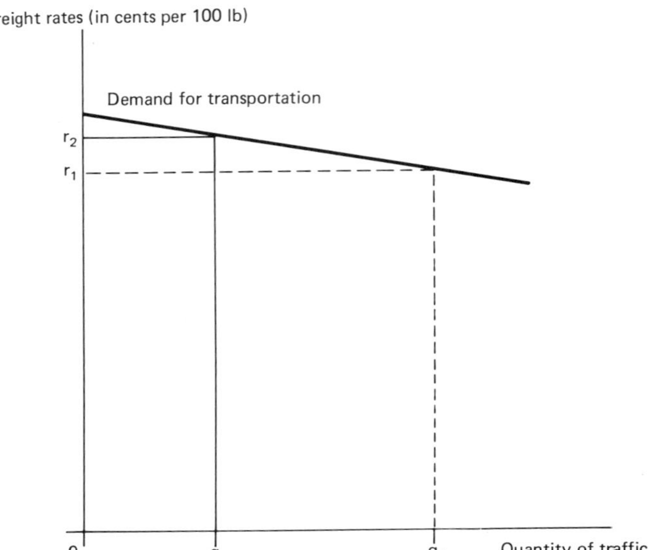

FIGURE 4-2
Freight revenue changes with elastic demand

gum, demand will be highly elastic due to the availability of many substitutes for bubble gum on the market. This is also true among the modes of transportation. If a particular product can only be carried efficiently by one mode (e.g., petroleum in a pipeline), the demand for that transportation service will be highly inelastic.

The foregoing discussion leads to this conclusion: Increased revenues will result from increased rates if the elasticity of demand is inelastic, since the revenue from the higher rates overbalances the decrease in revenue from the loss of traffic. If a logistics manager knows the elasticity of demand for the products the company produces, a more intelligent argument can be offered for or against rate increases. Although this line of thinking has been applied here to motor carriers, the same reasoning can be used for any of the other modes of transportation.

Innovations in the Motor Carrier Industry

The motor carrier industry has not been slow to adopt innovations to improve the quality and quantity of service. Specialized equipment is being developed, methods of speeding up loading and unloading are being installed at terminals, and the computerized location of shipments is aiding in the location and tracing of shipments.

Oil Pipelines

Oil pipelines are of two major types:

- Gathering lines, which haul the crude product from the wells to a concentration point.
- Trunk lines, which carry the crude product from concentration points to the refineries and market centers.

Not all pipelines, of course, carry petroleum products. Slurry pipelines, for example, carry a pulverized product (such as coal or grain) suspended in water, while the natural gas pipelines supply our furnaces and air conditioners.

Pipelines are not especially fast and, in fact, rank last in speed of the major modes of transportation since they move their contents at less than 5 miles per hour. To a degree, the capacity of a pipeline presents a problem to potential shippers. A line cannot accept economically a few hundred gallons of a product for shipment. Economical operation demands that some large minimum amount be offered for shipment, usually 25,000 barrels. The entire minimum tender of 25,000 barrels need not be delivered to the pipeline all at once, but the pipeline operator must have the assurance that the product will be offered for shipment within a reasonable time.

One of the most vexing problems in pipeline transportation is shrinkage of the product, due mostly to evaporation. Crude petroleum can lose up to one-quarter of a percent of its volume in transit through the line, while a refined product such as gasoline or kerosene can lose up to 1 percent. Another problem is the sheer size of the investment required to build the line. The 789 miles of the Alaskan pipeline, which began operations in 1977, cost $7.7 billion to build. Even acquisition of the right of way for the line can present problems. A proposed line to carry the Alaskan crude oil to the eastern seaboard was eventually abandoned owing to the difficulty experienced in getting permits from the many political jurisdictions along the proposed route.

Water Transportation

Water transportation within the nation travels along the rivers and canals, including the Great Lakes and the St. Lawrence Seaway. Speed is slow and made even slower when ice or floods clog the waterways; but water can carry large bulky cargoes very cheaply. This bulk cargo consists mainly of coal, petroleum, grain, and iron ore. The capacity of some modes of water transportation is quite large. A single modern towboat can now push 50 to 60 heavily laden barges at one time, while the vessels plowing the waters of the Great Lakes are regular deep-draft ships that could sail the oceans.

Weather presents a continuing problem. Ice can bring shipping to a complete halt, thus requiring companies that rely on water transportation to maintain enough inventory on hand to last through the icing season. Storms on the Great Lakes can be sudden and severe enough, with savage winds and high waves, to sink heavily laden vessels.

Perhaps the greatest technological advances have been in barge transportation. Special barges can be enclosed and equipped with the means of regulating both temperature and humidity. The LASH (lighter aboard ship) vessel is another example of these technical developments. Barges can be loaded on board ocean-going LASH vessels and then off-loaded in the water of the destination harbor to be towed or pushed to the unloading point in port.

Air Transportation

Air freight has never enjoyed the glamour accorded passenger aircraft, although many passenger airplanes are able to handle a considerable amount of freight in the cargo holds beneath the passenger compartments. Many jet aircraft can function in a dual capacity.

Airlines are classified into categories according to the size of their annual operating revenues:

- Major airlines earn operating revenues in excess of $1 billion annually.
- National airlines earn annual operating revenues between $75 million and $1 billion.
- Regional airlines earn operating revenues each year between $10 and $75 million.
- Medium regional airlines earn operating revenues of $10 million or less each year.

Air, by whatever type of airline, is generally considered a premium means of transportation. Air transportation is fast, but the freight rates are correspondingly high. Unless speed is an important factor in delivering parts to prevent the shutdown of an assembly line or to meet a delivery date to a valued customer, some less expensive means of transportation will usually suffice.

Probably the biggest headache faced by airline management is control over costs. The price of fuel skips up and down as the price of Mideast oil fluctuates. Deregulation has not improved the cost-control problem. Every time rates and fares are reduced to meet the competition, revenues are imperiled. In the days before the Airline Deregulation Act of 1978, everyone charged the same for travel between the same points, thus making revenues reasonably predictable. Now competition is unrestricted, and the delight of the consumer over lower charges is matched by the gloom of the airline executives who must now reduce operating costs in step with falling revenues.

Technological developments include new cargo-handling equipment at air terminals and the use of larger containers aboard the aircraft in which freight is shipped. As mentioned before, dual-purpose aircraft are in use in which cargo and passenger space can be converted one into the other and back again quite rapidly.

Other Modes of Transportation

Other modes of transportation include the freight forwarder who accepts small shipments and charges less carload (or less truckload or less planeload) rates and consolidates the small shipments into carload, truckload, or planeload lots, which are then sent by the lower quantity rates. Frequently, the freight forwarder acts as a traffic department for small companies, which usually ship in less carload lots. Shippers' cooperatives offer much the same service as the forwarder in consolidating small shipments into larger ones, except that the profits of the business are returned to members of the cooperative. Small packages can also be sent via parcel post, and can use some of the expedited delivery services the U.S. Postal System now offers. United Parcel Service (which demonstrated to the chagrin of the postal service that package mail could be handled at a profit) is another alternative, as is the package service offered by Greyhound and Trailways buses.

INTERMODAL TRANSPORTATION

One of the best-known examples of cooperation between two different modes of transportation is TOFC (trailer on flatcar), usually referred to as "piggy-back." Several different varieties of this service involving the shipment of loaded motor carrier trailers on a railroad flatcar exist, each providing different combinations of this combined rail and motor carrier service:

Plan I: The railroad provides only line-haul transportation of the trailer; the trailer itself is provided by a common motor freight carrier, which charges truck rather than rail rates for the service.

Plan II: The railroad handles the entire move door to door using its own trailers and flatcars.

Plan II½: The railroad provides the trailer and flatcar, but the shipper transports the trailer to the rail terminal, while the consignee at the destination transports the trailer from the freight station to the place of business for unloading.

Plan III: The railroad provides only the flatcar and charges a flat fee rather than a mileage rate; the shipper or a freight forwarder provides the trailer.

Plan IV: The railroad provides only the locomotive that pulls the train; the shipper furnishes both the trailer and the flatcar and is responsible for loading and unloading the flatcar.

Plan V: Joint rail–truck rates are charged for the shipment; the equipment is provided by common rail and motor carriers.

Container service, in which the cargo is first loaded into large steel boxes that can be locked, is becoming a standard method of shipping small packages in large quantities by any mode of transportation. Container on flatcar (COFC) is available, but the service began in the 1940s in ocean shipping and still predominates there, where the use of these containers has reduced stevedoring costs by as much as 90 percent.

The *bridges* in transportation involve water–rail–truck movements using both national and international means of transportation. The *minibridge* moves a container from a foreign port, through an intermediate U.S. port, to a final destination U.S. port, as portrayed in Figure 4-3.

The *microbridge,* as illustrated in Figure 4-4, moves the container from a foreign port, through a U.S. port, and then overland to a U.S. city.

The *land bridge,* portrayed in Figure 4-5, moves the container between foreign ports by traversing the length of the United States between oceans.

Even from this brief overview of the transportation system, the conclusion is easily reached that a wide variety of modes of shipment is available to shipping companies. So vast and complex is the system, however, that much managerial effort is required to ensure obtaining the most efficient low-cost service desired. This is where the special skills of the traffic manager become an essential part of the overall problem of logistics management.

OBJECTIVES ACHIEVEMENT CHECKUP

The following exercises will help you in assessing the extent to which you have achieved the objectives for this chapter. Where essay-type answers are required, the more you are able to condense your knowledge into a short paragraph of a

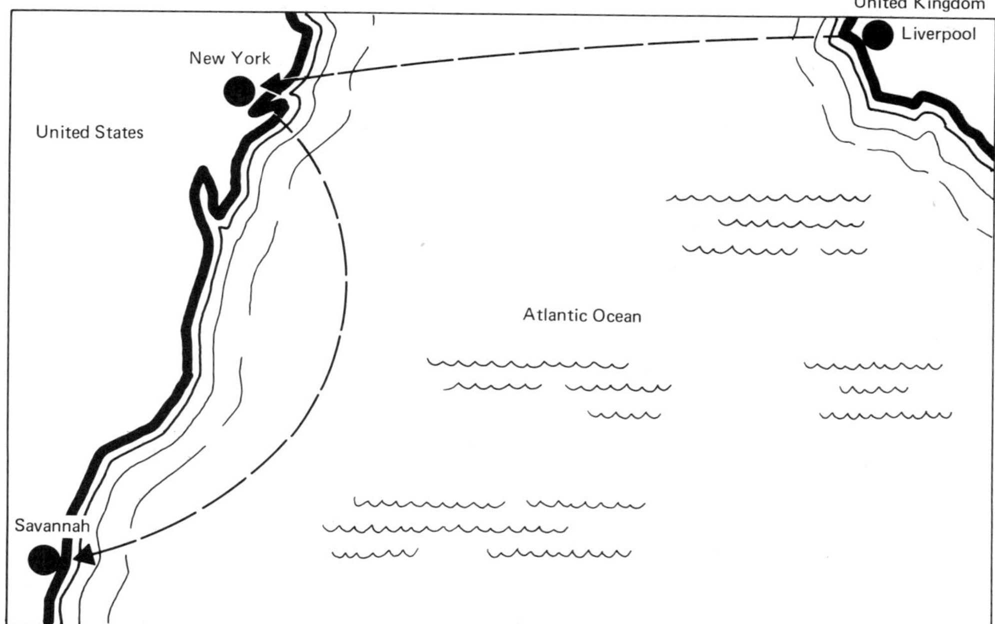

FIGURE 4-3
Minibridge

few sentences the better. Looking back as you work these exercises and rereading parts of the chapter are permissible; but do not look ahead to the answers until you have done your best with the questions.

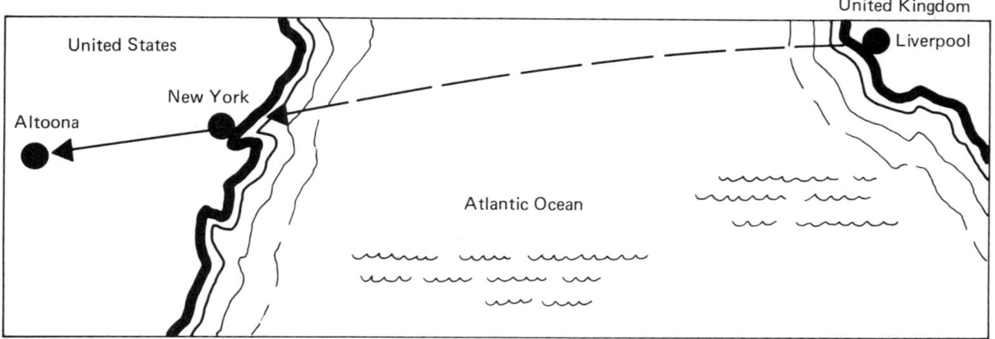

FIGURE 4-4
Microbridge

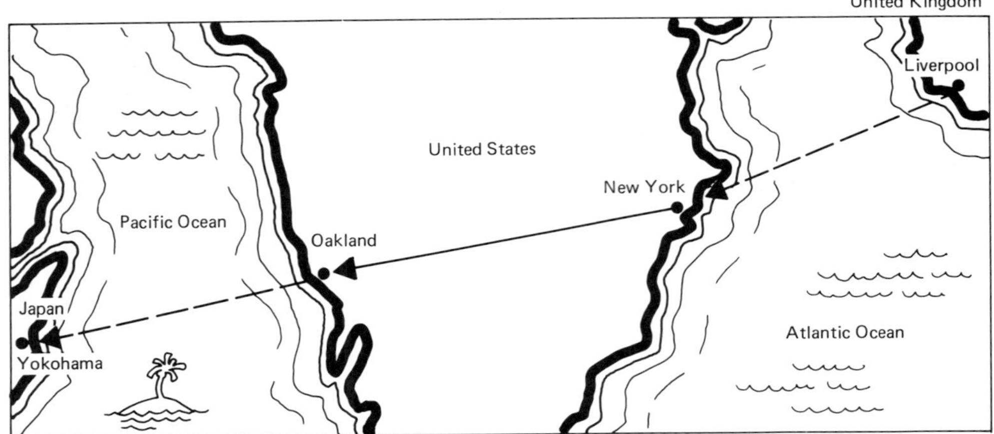

FIGURE 4-5
Land bridge

1. What would cities look like today without modern means of transportation? Let your imagination help in answering this question.
2. What is the one most significant advantage and disadvantage of each of the major modes of transportation?
3. How does transportation support the logistical function?
4. The transportation bridges are placed where in the logistical chain?
5. List some examples of intermodal service.

ANSWERS TO CHECKUP QUESTIONS

1. Without modern methods of transportation, cities would probably be much smaller and self-sufficient to the maximum extent; farm areas that provide food would be much closer in; and residents would live closer to their places of employment.

2.

Mode	*Advantage*	*Disadvantage*
Rail	High capacity	Poor loss and damage record
Water	Low cost	Poor dependability
Air	Speed	High cost
Highway	Speed and low cost	Low capacity
Pipeline	Dependability	Slow speed

3. Transportation supports company logistical functions in the following ways:

a. Speed affects the size of inventory stocks.
b. Availability affects the location of plants and storage facilities.
c. Handling en route determines packaging requirements.
d. Carrier equipment affects the type of materials-handling equipment needed.
e. Reliability affects achievement of customer service goals.
f. Rates and services, as part of the costs of production, influence the price charged for the product.

4. The transportation bridges are:
   a. Minibridge: Freight (usually packed in a steel container box) moves from a foreign port, through a U.S. port, to another U.S. port.
   b. Microbridge: Freight moves from a foreign port through a U.S. port overland to a U.S. city.
   c. Land bridge: Freight moves from a foreign port, overland across the United States, to another foreign port.
5. Examples of intermodal service are:
   a. Trailer on flatcar (TOFC) or "piggy-back."
   b. Container service in which the steel box is carried by a variety of modes before reaching a final destination.

5

Using the System of Transportation

CHAPTER OBJECTIVES

Upon completion of this chapter, you should be able to:

A. Trace the origins of regulation in the transportation system.
B. Understand the legal obligations of carriers.
C. Explain the ways in which freight rates are structured.
D. Calculate the cost of a shipment using class rates.
E. Distinguish between the various types of bills of lading.

Freight rate tariffs, the price lists for transportation services published by carriers, are truly marvels of complexity. The beholder of these wondrous examples of human ingenuity and the publisher's art are constrained to wonder where they all came from. The rites of passage from the simple lists of freight charges in the 1840s to the modern rate tariffs were not easy, and the guide posts were court cases and legislative enactments.

LEGISLATIVE BACKGROUND

Since the Airline Deregulation Act of 1978, the transportation industry has been moving from a highly regulated environment to one in which competitive forces are given freer reign. The partial deregulation of motor carriers in 1980, for example, permitted the granting of discounts on volume shipments, while the Staggers Rail Act of 1980 permitted railroads to enter into contracts with shippers for the first time.

The basis for regulation was established in 1670. In that year, England's Lord Chief Justice Hale, in his essay *De Pertibus Maris,* held that public facilities like ports must be controlled to protect the public good even though privately owned. This English doctrine was used after the Civil War in the United States to moderate the outrageous conduct of the railroads toward the public they were supposed to serve.

First, however, the public utility concept had to be incorporated into American common law. The landmark case concerned grain elevators. Munn and Scott, operators of a grain elevator in Chicago, refused to obtain the license required by Illinois state law or observe the restrictions on the maximum rates that could be charged. They were hauled into court (although the license was only $100), but appealed when the state won. The case reached the Supreme Court as *Munn* v. *Illinois (1877),* and once again Illinois won, thereby establishing the right of public authorities to regulate private businesses "affected with a public interest." Illinois then promptly applied this doctrine to the regulation of the rates charged by railroads on freight moving within and through the state. The case also established another point of law: Grain, or any other kind of freight, moving into or out of the state moves in interstate commerce. Although the regulation of interstate commerce is a responsibility of the federal government, the states could do so if the federal government had not yet acted.

Less than a decade later, the whole situation fell apart. The Wabash Railroad was caught charging more for hauls between Gilman, Illinois, and New York City than between Peoria, Illinois, and New York.

The trouble with this arrangement was the fact that Gilman was 86 miles *closer* to New York than Peoria. The state of Illinois informed the railroad that it was not going to let the carrier charge more for short hauls than for long hauls along the same route. The railroad appealed, and the *Wabash Case (1886)* became another landmark decision: The railroad won this one and *Munn* v. *Illinois* was reversed. The Supreme Court agreed with the Wabash Railroad: Only the federal government could regulate interstate commerce and the states could not under the Constitution, even though the federal government had not yet acted.

In this case, the federal government had not yet acted, but soon did. With about 80 percent of all freight movements now subject to no control whatever, Congress moved with commendable speed and passed the Act to Regulate Commerce (now known as the Interstate Commerce Act) the following year in 1887. This law established the Interstate Commerce Commission to administer the law, and government regulation of transportation was on its way!

Other modes of transportation were soon brought under the federal regulatory umbrella. Oil pipelines were brought under regulation by the Hepburn Act of 1906; motor carriers had their turn in the Motor Carrier Act of 1935; airlines came under the Civil Aeronautics Board rather than the ICC by the Civil Aeronautics Act of 1938; domestic water transportation was controlled by the Transportation Act of 1940 with the Federal Maritime Commission in charge; and the last bastion of unregulated transportation—the freight forwarder—fell under control of the ICC by the Freight Forwarder Act of 1942.

Beginning in 1977 with the deregulation of air freight, the whole system of commission control has been changing. In general, the deregulation of air, rail, and motor freight carriers has made entry into each industry much easier. Witness the surge in new airlines created after 1978. Rate negotiations between carriers and shippers have been made easier and less formal, while service in terms of routes and schedules has been made more flexible.

WHAT MUST A CARRIER DO?

Common carriers must serve all who request their services, provided (1) the carrier has the route authority to haul the freight from origin to destination and (2) possesses the requisite equipment. A dairy farmer, for example, with a large quantity of milk cannot force the shipment on a carrier that does not have tank trucks and never has held itself out as a hauler of bulk liquids. A carrier cannot, however, refuse a small shipment simply because the revenue from the shipment does not cover the cost of carriage. In contrast, a common carrier can refuse a shipment by imposing an embargo. This involves a refusal to accept a shipment for a specific destination because any more freight sent to that destination would only make a congested situation worse.

The common carrier is responsible for the safe delivery of freight, which means the carrier is liable for loss or damage to the cargo. Exceptions to this rule of liability, however, are numerous, such as:

- Acts of God, which would include natural disasters such as floods, forest fires, hailstorms, and tornadoes.
- "Enemies of the King," which means damage or loss due to war, either by uniformed combatants or saboteurs.
- Acts of public authorities, which means an overlong inspection, for example, by state authorities, which results in the spoilage of the product.
- Acts of the shipper through improper packaging or loading of the product.
- Inherent nature of the product, such as the evaporation of liquids.

The carrier must charge reasonable rates and not discriminate by customer, type of product being shipped, or geographical location.

The contract carrier has rights, duties, and liabilities specified by a contract between the carrier and the shipper. Private carriers are not usually involved in disputes concerning what they are legally obligated to do since this type of carrier owns both the transportation equipment and the product hauled. At one time, carriers tried to evade regulation by buying the products from shippers and then selling it at destination for a price that just covered the cost of purchasing the goods and the transportation. This is now quite illegal.

Exempt carriers could be any type of carrier, but it gains its exemption from regulation by the kind of product hauled. These carriers negotiate rates directly with shippers since they are exempt from rate control. Any carrier becomes exempt when it carries:

- Unprocessed farm products in trucks
- Bulk products and liquids in barges
- Livestock
- Fish

Finally, all carriers must observe federal, state, and local regulations concerning speed and weight limits and safety.

UNDERSTANDING FREIGHT RATES

Despite the laws that seem to control in great detail the establishment of freight rates, most rates are not set by regulatory agencies. Instead, most freight rates are set by direct contract between shippers and carriers, between carriers and rate bureaus, or between the carriers themselves.

Three economic bases are possible as guidelines in setting the prices for transportation. *Full-cost pricing* requires that the rate cover all costs incurred in providing the service, to include:

- Fixed costs: costs that do not change as the amount of traffic changes, such as maintenance of track and roadbed, terminals, equipment, bridges, and signals.
- Variable costs: costs that do change as the volume of traffic changes.
- A fair margin of profit.

Value-of-the-service pricing sets variable costs as a floor below which rates cannot be lowered. Since variable costs represent the out-of-pocket costs that keep the carrier operating hauling freight, these expenses must be met as an absolute minimum. Fixed costs, on the other hand, have already been incurred when the line was built (for example, how many times does a freight depot have to be rebuilt?) and, in the short run, do not need to be covered. In fact, a large transportation company like a major railroad can stay in business for years without making a profit as long as the operating expenses are covered. Ultimately all costs, both fixed and variable, must be recovered; otherwise, bankruptcy ensues.

Marginal cost pricing is probably the most difficult to understand. In this type of pricing, only the additional cost of providing an extra unit of service needs to be covered by the rate charged. For example, how many costs should be covered by the fares charged when an extra jetliner is placed in service by an airline between two points on an established route? The terminals have already been built, the ticket counters are functioning, and the baggage-handling facilities are already in place. The cost of flying an additional airplane does not require the expansion of existing facilities; therefore, the only costs that need to be included in the fares charged are the fuel and lubricants consumed in operating the aircraft, in-flight meals, flight-crew wages, and other costs connected only and directly with flying that one planeload of passengers and freight. The same is true in any other mode of transportation. How many additional costs are incurred by coupling one more boxcar into a train? These additional costs are the marginal costs that are relevant in rate setting.

Rate Structures

Rates can be organized, or structured, in several different ways. One of the most frequently encountered rate structure is the *volume-related rate.* These are the types of rates that offer a lower price for carrying full carloads, truckloads, or planeloads of freight, but a higher price for hauling smaller less-than-carload shipments. This type of volume discount is justified to cover the additional handling costs incurred in dealing with small shipments and the savings that result in handling the larger shipments.

Distance-related rates increase in the same proportion as distance increases. In other words, if the distance the freight is hauled doubles, the rate also doubles. *Tapered rates* are a variation on distance rates. The tapered rate increases as distance increases, but in lesser proportion. Thus, should the distance double, the rate might only increase by one-third. For example, a tapered rate might change from $1.14 per hundred pounds at a distance of 100 miles to only $1.49 per hundred pounds when the distance reaches 200 miles, or an increase of only one-third in the rate even though the distance has doubled.

Differential rates are found by adding or subtracting a specific amount—the differential—from some standard rate.

Types of Freight Rates

Class rates are a standard price for transportation based on a rating, or class, to which the product is assigned. If no other special rate can be found applicable to a particular product, then the freight moves on class rates, which are usually the highest prices that the carrier can charge for its services. Keep in mind, the difference between a rate and a rating—a difference often confused. A *rate* is a money price for hauling freight, usually expressed as so many cents per hundred pounds (often abbreviated as cwt in tariff publications); a *rating* is a number indicating a particular class of products to which the product is assigned. The rating, of course, is useful in determining the rate or price.

Rail rate tariffs contain 31 classes, or ratings, ranging from a high of 400 to 13. One extremely convenient aspect in finding the exact rate on which a shipment should move is the use of class 100. Not every tariff lists the rates connected with all 31 classes; but even though a particular class rating is not included, knowledge of the rating and the class 100 rate permits calculating the desired money price. The money price, or rate, for class 45, for example, is simply 45 percent of the class

100 rate. Ratings for motor carriers range from 500 to 35, for a total of 23 classes. The rates for motor carriers are similarly derived from these class ratings.

An *exceptions rating* places a commodity in a class other than the one to which normally assigned. The exceptions rate that results from this practice is usually lower and frequently also involves a lower minimum weight or lighter packaging requirements.

A *commodity rate* is a specific money price applicable to a specific product moving between specified points. Almost three-quarters of all rail freight shipments move on these special commodity rates. For the three rates mentioned thus far—the class rate, exceptions rate, and commodity rate—the commodity rate is senior and takes precedence over the other two. In the event all three types of rates for a single product exist, the commodity rate will be charged. In contrast, if neither an exceptions rate nor a commodity rate can be found covering a particular shipment, then the last resort is the class rate.

Several special types of rates also exist. The *incentive rate* is used to encourage shippers to pack more freight in a car and in so doing use the equipment more intensively. The *volume rate* provides a lower rate if the shipment exceeds the carload or truckload minimum weight. Low rates are charged if a shipper can use 2 to 20 cars at one time. This *mulitple car rate* simplifies the problem of providing equipment since cars can be assembled in batches. The *trainload rate* would apply if over 20 cars are used that can be coupled into a single train. Coal is frequently shipped in these unit trains with nothing between the locomotive and the caboose but up to 100 hopper cars.

A *local rate* will be charged if the entire shipment is carried between two points on the lines of the same carrier, while a *joint rate* would apply on shipments that originate on the lines of one carrier and end at a destination on the line of a different carrier. A *through rate* is a single rate applying on a shipment from origin to destination. The *combination rate* is the sum of local or joint rates charged on a shipment from origin to destination when no single joint or through rate exists.

One of the most useful of the special types of rates is the *all-commodity rate,* sometimes called FAK or "freight all kinds." This rate is quoted only in terms of weight, not type of product. A shipper could use this rate to ship a number of different products in a single freight car as long as the weight requirement was met for the total shipment. In this way, a shipper could qualify for the lower carload rate even though the amount of each separate commodity fell short of the minimum weight requirement for carload shipments. A *released value rate* is a special low rate charged if the shipper will agree to reduced carrier liability for loss or damage to the shipment.

Blanket or *group rates* are of three varieties:

- Point to area, which applies to shipments, as an example, from Chicago to points in the New England area.
- Area to point, which might apply to shipments moving from the Pacific Coast states to Omaha.
- Area to area, which might apply from points in the Pacific Coast states to points in New England.

Section 22 rates (referring to Section 22, now Section 10721, of the Interstate Commerce Act) are special rates that apply on shipments of government freight. *Export–import rates* apply on shipments with overseas points as origin or destination.

The continuing trend toward deregulation in the transportation industry should spawn many innovative rate proposals. The preceding list, though, should provide a good idea of how complex the whole area of freight rates has become. Add to this the variety of passenger fares in surface and air modes of transportation and the complexity is compounded. At one time, professionals in the industry thought impossible the computerizing of trillions of individual rates quoted in over 200,000 tariff publications now in use. The situation is being improved by simplifying the descriptions of commodities and the route listings.

Many companies have found a partial solution to the problem of finding the lowest legal rate for their shipments by computerizing at least the rates along the routes customarily used for their products. The sheer complexity of rates virtually demands that all companies audit their freight bills either by an auditor in the employ of the firm or by hiring a free-lance auditor. Freight bill audits frequently return more than their cost in reimbursement from carriers for inadvertent overcharges.

FINDING RATES

To get some idea of what is involved in computing freight rates, let's go through the procedure for finding a class rate. First, the freight must be classified. The basic idea of classification is simple enough: Group a large number of commodities with related characteristics into a few classes. Second, assign each class a rating number and use this to calculate the money rate in the customary cents per 100 pounds. This money rate will be based both on the nature of the commodity being hauled and the distance traveled.

Note the distinction: A *rating* is a number indicating the class into which an item has been placed by the process of classification; a *rate* is a money price determined in part by the rating. The other determinant of the money price is distance and is established by the *rate basis number*. The rate basis number reflects the distance the shipment will move and is set by the points of origin and destination. The procedure for finding a rate thus is rather clear:

- First, locate the item to be shipped in a freight classification tariff to determine the rating. Two ratings will be found: one for carload shipments and another for less carload shipments. The minimum weight that qualifies a shipment as a carload will also be shown.
- Second, locate the rate basis number in a freight rate tariff that applies between the points of origin and destination.
- Third, use the rating and the rate basis number to find the money rate.

Suppose a company wanted to ship 22,000 pounds of clotheslines from Brookneal, Virginia, to Ansonia, Pennsylvania. Following our procedure, we would first turn to a freight classification tariff and locate the item. In the tariff extract shown in Table 5-1, clotheslines are Item 28480. The carload minimum weight column tells us that 24,000 pounds (the R following the weight figure indicates that the railroad can offer railcars of varying types that might take a different minimum weight) is required to qualify the shipment as a carload. Were that the case here, the carload rating would be 55. Unfortunately, this shipment contains only 22,000 pounds; therefore, the less carload rating of 85 must be used. Don't despair; we may yet be able to use that carload rating.

Now we turn to a freight tariff to find the rate basis number. Table 5-2 shows a page of rate basis numbers covering the cities between which the clotheslines will travel. The rate basis applicable between Brookneal and Ansonia is 488.

The last step is to find the money rate. For this, we turn to the page from a rate tariff in Table 5-3. Rate basis number 488 falls in the group "481 to 500," which will be used to find the exact rate. Moving across the line containing the rate basis number, we find a rate of 196 cents per 100 pounds.

Calculation of the freight bill for this shipment is now easy. Multiply the rate by the weight:

$$\$1.96 \times 220 = \$431.20$$

Are you puzzled by the "220" figure? Recall the rate is in cents per 100 pounds; therefore, it is necessary to reduce the 22,000 to hundreds, or 220.

TABLE 5-1
Partial Page from the Uniform Freight Classification

UNIFORM FREIGHT CLASSIFICATION 10 — 28275-28511

ITEM	*ARTICLES*	*Less Carload Ratings*	*Carload Minimum (Pounds)*	*Carload Ratings*
	CLOTH, DRY GOODS OR FABRICS (Subject to Item 27730)—Concluded:			
28275	Pile fabric, plush, velour, velvet or velveteen, cotton, mohair, rayon, wool or synthetic fibre, separate or combined:			
	With cellular, expanded or foamed plastic or cellular expanded or foamed rubber backing, in packages	150	10,000R	85
	Without cellular, expanded or foamed plastic or cellular expanded or foamed rubber backing			
	On frames, creels or racks, in boxes	150	10,000R	85
	Not on frames, creels or racks, in packages	100	20,000R	70
28284	Ramie cloth, in bales or rolls	55	AQ	55
28287	Ribbon bows or rosettes, with or without decorations of same or other materials, in packages	250	AQ	250
28290	Roofing, cotton duck (cotton canvas), waterproofed, coated with asphalt, pitch, tar or similar materials, or waterproofed by saturation, but not combined with felt or paper, in bales, boxes, crates or rolls.	85	30,000	55
28300	Roofing, glass fibre, bitumen impregnated, in boxes, crates or rolls	100	30,000	70
28310	Sheets, cotton or tobacco picking, in machine pressed bales	55	30,000	$37\frac{1}{2}$
28320	Sheets, tobacco shipping, burlap, gunny, ixtle (istle) jute or sisal:			
	In packages other than machine pressed bales	65	30,000	$37\frac{1}{2}$
	In machine pressed bales	55	30,000	$37\frac{1}{2}$

TABLE 5-1
Partial Page from the Uniform Freight Classification (*cont.*)

UNIFORM FREIGHT CLASSIFICATION 10 — 28275-28511

ITEM	ARTICLES	Less Carload Ratings	Carload Minimum (Pounds)	Carload Ratings
28350	Tire fabric, tire cord, hose cord, belting cord, hose yarn, hose fabric, or belting fabric, LCL, in bags, barrels, boxes or wrapped bales or rolls; CL, in packages	55	30,000	40
28360	Webbing, burlap or jute, in bales or boxes.	70	30,000	40
28390	Window shade cloth or felt, noibn, in boxes	100	20,000R	70
28405	Cloth or tape, insulating, glass fibre, in boxes	100	20,000R	70
28410	Cloth or tape, insulating, noibn, in packages	$77\frac{1}{2}$	30,000	45
28420	Cloth winding boards:			
28430	Corrugated fibreboard, paper covered, or solid fibreboard with built-up ends, in packages	70	24,000R	45
28440	Solid fibreboard, noibn, in packages	65	36,000	35
28450	Wooden, solid, in packages.	55	36,000	35
28460	Cloth winding frames, fibreboard or wood, with or without paper covers, in packages	100	14,000R	70
28470	Clothes driers or racks, steel, noibn, completely KD or spacers removed and sides collapsed, in boxes	70	30,000	40
28480	Clotheslines in holders, in barrels or boxes .	85	24,000R	55
28490	CLOTHING:			
28500	Belts, noibn, in barrels or boxes	100	20,000R	70
28510	Clothing, fur or fur lined, noibn, see Note 2, Item 28511, in boxes, or in salesmen's trunks, locked and sealed with metal seals:			

For explanation of abbreviations, numbers and reference marks, see last page of this Classification; for packages, see pages following rating section.

Source: Reprinted from Ronald H. Ballou, *Basic Business Logistics* (Englewood Cliffs, N.J.: Prentice-Hall Inc., 1978), Table 7-2, p. 170.

TABLE 5-2
Table of Selected Rate Base Numbers

FREIGHT TARIFF No. 1009A

SECTION 1
APPLICATION OF RATE BASES

BETWEEN ☛ (See item 100)

AND (See Item 100)

RATE BASES APPLICABLE (For rates, see pages 282 to 337)

AND ↓ / BETWEEN →	Annapolis . . . Md. (See Note 9)	Ann arbor . . Mich.	Ansonia . . . . Pa.	Anthony . . . W. Va.	Arcade . . . . N.Y.	Arlington . . . . Vt.	Ashley . . . . Mich.	Ashtabula . . . Ohio	Athens . . . . Ohio	Atlantic City . . N.J.	Augusta . . . . Me.	Aurora . . . . . Ill.	Avondale . . . . Pa.	Bad Axe . . . Mich.	Baltimore . . . Md.	Bangor . . . . . Me.
Annapolis (See Note 9) . . . . . . Md.	40															
Ann Arbor . . . . Mich.	629	40														
Ansonia . . . . Pa.	291	418	40													
Anthony . . . . W. Va.	350	494	438	40												
Arcade . . . . N.Y.	404	304	114	523	40											
Arlington . . . . Vt.	405	597	302	653	330	40										
Ashley . . . . Mich.	712	84	457	577	343	632	40									
Ashtabula . . . . Ohio	465	207	232	441	157	452	290	40								
Athens . . . . Ohio	439	245	410	252	375	677	328	226	40							
Atlantic City . . . . N.J.	205	700	317	463	430	329	765	510	549	40						
Augusta . . . . Me.	610	874	571	882	607	315	909	729	954	530	40					
Aurora . . . . Ill.	836	267	651	637	570	863	285	420	404	907	1140	①				
Avondale . . . . Pa.	115	606	238	371	351	316	689	415	453	107	521	813	40			
Bad Axe . . . . Mich.	702	146	412	624	298	587	120	336	375	720	864	366	649	40		
Baltimore . . . . Md.	40	593	255	314	368	383	676	429	403	169	588	800	79	666	40	
Bangor . . . . Me.	685	949	654	957	682	390	984	804	1029	605	74	1215	596	930	663	40
Barry . . . . Ill.	978	481	840	752	766	1061	507	609	539	1075	1338	①	981	588	942	1413
Barstow . . . . Ill.	954	384	769	744	688	981	391	537	519	1025	1258	①	931	484	918	1333
Batavia . . . . N.Y.	418	304	142	559	40	293	343	159	384	426	570	570	365	298	382	645
Bath . . . . N.Y.	340	378	63	521	115	284	417	237	455	348	561	644	287	372	304	636
Battle Creek . . . . Mich.	696	89	501	561	387	676	93	273	312	767	953	193	673	174	660	1028
Beckley . . . . W. Va.	414	437	550	91	569	718	520	419	195	527	946	585	435	567	378	957
Bedford . . . . Pa.	238	436	180	278	255	449	519	265	278	322	725	643	228	546	202	800
Bellaire . . . . Mich.	875	252	585	746	471	760	183	458	497	893	1037	339	822	216	839	1112
Bellefontaine . . . . Ohio	568	142	425	337	351	646	225	194	129	658	923	278	565	271	532	998
Bellefonte . . . . Pa.	262	444	90	372	184	359	527	248	351	286	636	651	207	432	226	711
Bellows Falls . . . . Vt.	454	678	380	726	411	93	713	533	758	374	241	944	365	668	432	316
Bennington . . . . N.H.	463	701	406	735	434	188	736	556	781	383	200	967	374	691	441	275
Bens Run . . . . W. Va.	399	310	336	288	344	633	393	209	75	507	910	479	413	439	363	985
Benton Harbor . . . . Mich.	754	165	569	584	469	760	164	353	351	825	1037	128	731	258	718	1112
Berlin . . . . N.H.	593	805	519	865	537	232	838	660	885	513	133	1070	504	793	571	179
Berwick . . . . Pa.	224	537	129	408	242	281	585	337	457	223	556	757	163	540	188	631
Bethel . . . . Me.	621	843	563	893	576	266	877	699	924	541	105	1109	532	832	599	152
Bethel . . . . Vt.	518	742	444	790	475	157	777	597	822	438	237	1008	429	732	496	512
Bethlehem . . . . Pa.	197	611	211	416	324	240	650	419	476	115	482	828	91	605	161	557
Big Blizzard Run . . . . W. Va.	454	483	463	275	476	729	566	382	252	562	969	655	468	612	418	1044
Binghamton . . . . N.Y.	323	467	113	517	195	189	506	322	517	269	466	733	230	461	287	541
Birmingham . . . . N.J.	174	666	287	434	400	366	722	475	513	75	468	873	73	677	138	533
Blackstone . . . . Va.	262	688	462	222	575	589	771	600	446	375	794	829	283	813	226	869
Blanchester . . . . Ohio	557	244	480	336	425	720	327	268	118	665	997	333	571	374	521	1072
Bloomington . . . . Ill.	851	330	690	626	615	910	348	458	412	941	1187	①	847	430	815	1262
Bloomington . . . . Ind.	735	300	622	494	547	842	357	390	296	832	1119	232	738	437	699	1194
Bluefield . . . . W. Va.	412	471	570	162	616	739	554	466	242	525	944	607	433	601	376	1019
Bluffton . . . . Ind.	661	149	487	461	413	708	206	256	222	737	985	195	643	289	625	1060
Boston . . . . Mass.	450	743	448	722	476	184	778	598	794	370	171	1009	361	733	428	246
Boyne City . . . . Mich.	868	245	578	739	464	753	183	451	490	886	1030	362	815	209	832	1105
Branchland . . . . W. Va.	532	339	526	211	496	798	422	347	123	645	1064	475	553	469	496	1139

TABLE 5-2
Table of Selected Rate Base Numbers (*cont.*)

FREIGHT TARIFF No. 1009A

SECTION 1
APPLICATION OF RATE BASES

BETWEEN (See item 100) / AND (See Item 100)	*Annapolis . . . Md. (See Note 9)*	*Ann arbor . . . Mich.*	*Ansonia . . . . . . Pa.*	*Anthony . . . W. Va.*	*Arcade . . . . . . N.Y.*	*Arlington . . . . . Vt.*	*Ashley . . . . . . Mich.*	*Ashtabula . . . Ohio*	*Athens . . . . . Ohio*	*Atlantic City . . N.J.*	*Augusta . . . . . . Me.*	*Aurora . . . . Ill.*	*Avondale . . . . . Pa.*	*Bad Axe . . . . Mich.*	*Baltimore . . . . Md.*	*Bangor . . . . . . Me.*
	RATE BASES APPLICABLE (For rates, see pages 282 to 337)															
Brandywine . . . . Md.	95	599	295	293	408	422	682	442	408	209	627	806	116	706	59	702
Brave . . . . Pa.	362	381	300	263	313	582	464	240	198	453	859	588	376	510	326	934
Bremo . . . . Va.	215	665	415	200	527	540	748	532	423	327	747	808	236	795	179	822
Brewster . . . . N.Y.	293	655	337	565	388	148	691	510	637	213	343	921	204	646	271	418
Bridgeton . . . . N.J.	187	682	300	595	413	317	748	492	529	60	518	889	89	703	151	593
Bristol . . . . Va.-Tenn.	460	551	618	309	728	787	610	568	349	573	992	653	481	680	424	1067
Bristolville . . . . Ohio	439	221	266	407	191	486	304	45	197	509	763	433	415	350	403	838
Brooklyn . . . . Wis.	933	322	713	737	599	888	273	516	500	1004	1165	(1)	910	375	897	1240
Brookneal . . . . Va.	290	641	488	175	598	617	784	571	399	403	822	776	311	771	254	897
Brownsville . . . . Pa.	322	334	255	280	268	537	417	177	215	430	814	541	336	461	286	889
Brunswick . . . . Me.	577	841	546	849	574	282	876	696	921	497	40	1107	488	831	555	107
Buffalo . . . . N.Y.	440	268	150	552	40	329	307	127	352	458	606	534	387	262	404	681
Burkeville . . . . Va.	254	673	454	207	567	581	756	592	431	367	786	814	275	803	218	861

For explanation of reference marks and notes, see pages 7 to 9.

Source: Reprinted from Ronald H. Ballou, *Basic Business Logistics* (Englewood Cliffs, N.J.: Prentice-Hall, Inc., 1978), Table 7-3, p. 173.

Managers are always fond of employees who can find ways to save the company money. Let's try. Would it be more advantageous to pay for an extra 2,000 pounds of "air" to qualify this shipment for the carload rate?

Look back to the freight classification page in Table 5-1 and find the carload rating for these clotheslines. The carload rating is 55. The rate basis number of 488 will not change since the points of origin and destination have not changed. When we look for a rate under class 55 in the freight tariff page in Table 5-3, a problem arises: The tariff does not contain a column for class 55.

Fortunately, the problem is easily resolved. Each class is simply a percentage of class 100. We can find the class 55 rate, then, by simply taking 55 percent of the class 100 rate.

The class 100 rate applicable to rate basis number 488 is 231 cents per 100 pounds. The rate for class 55 is 55 percent of this figure:

$$\$2.31 \times 55\% = \$1.2705$$

TABLE 5-3
Table of Rates in Cents per 100 Pounds from Freight Tariff No. 1009A

RATE BASES NUMBERS (NUMBERS INCLUSIVE)	CLASSES											
	400	300	250	200	175	150	125	110	100	$92\frac{1}{2}$	85	$77\frac{1}{2}$
40	344	258	215	172	151	129	108	95	86	80	73	67
41 to 45	356	267	223	178	156	134	111	98	89	82	76	69
40 to 50	364	273	228	182	159	137	114	100	91	84	77	71
51 to 55	376	282	235	188	165	141	118	103	94	87	80	73
56 to 60	384	288	240	192	168	144	120	106	96	89	82	74
61 to 65	396	297	248	198	173	149	124	109	99	92	84	77
66 to 70	404	303	253	202	177	152	126	111	101	93	86	78
71 to 75	412	309	258	206	180	155	129	113	103	95	88	80
76 to 80	424	318	265	212	186	159	133	117	106	98	90	82
81 to 85	432	324	270	216	189	162	135	119	108	100	92	81
86 to 90	440	330	275	220	193	165	138	121	110	102	94	85
91 to 95	448	336	280	224	196	168	140	123	112	104	95	87
96 to 100	456	342	285	228	200	171	143	125	114	105	97	88
101 to 110	472	354	295	236	207	177	148	130	118	109	100	91
111 to 120	488	366	305	244	214	183	153	134	122	113	104	95
121 to 130	504	378	315	252	221	189	158	139	126	117	107	98
131 to 140	516	387	323	258	226	194	161	142	129	119	110	100
141 to 150	532	399	333	266	233	200	166	146	133	123	113	103
151 to 160	544	408	340	272	238	204	170	150	136	126	116	105
161 to 170	560	420	350	280	245	210	175	154	140	130	119	109
171 to 180	572	429	358	286	250	215	179	157	143	132	122	111
181 to 190	584	438	365	292	256	219	183	161	146	135	124	113
191 to 200	596	447	373	298	261	224	186	164	149	138	127	115
201 to 210	612	459	383	306	268	230	191	168	153	142	130	119
211 to 220	624	468	390	312	273	234	195	172	156	144	133	121
221 to 230	636	477	398	318	278	239	199	175	159	147	135	123
231 to 240	648	486	405	324	284	243	203	178	162	150	138	126
241 to 260	672	504	420	336	294	252	210	185	168	155	143	130
261 to 280	692	519	433	346	303	260	216	190	173	160	147	134
281 to 300	716	537	448	358	313	269	224	197	179	166	152	139
301 to 320	740	555	463	370	324	278	231	204	185	171	157	143
321 to 340	760	570	475	380	333	285	238	209	190	176	162	147
341 to 360	784	588	490	392	343	294	245	216	196	181	167	152
361 to 380	804	603	503	402	352	302	251	221	201	186	171	156
381 to 400	824	618	515	412	361	309	258	227	206	191	175	160
401 to 420	844	633	528	422	369	317	264	232	211	195	179	164
421 to 440	864	648	540	432	378	324	270	238	216	200	184	167

TABLE 5-3
Table of Rates in Cents per 100 Pounds from Freight Tariff No. 1009A (*cont.*)

RATE BASES NUMBERS (NUMBERS INCLUSIVE)	CLASSES											
	400	300	250	200	175	150	125	110	100	92½	85	77½
441 to 460	884	663	553	442	387	332	276	243	221	204	186	171
461 to 480	904	678	565	452	396	339	283	249	226	209	192	175
481 to 500	924	693	578	462	404	347	289	254	231	214	196	179
501 to 520	944	708	590	472	413	354	295	260	236	218	201	183
521 to 540	964	723	603	482	422	362	301	265	241	223	205	187
541 to 560	980	735	613	490	429	368	306	270	245	227	208	190
561 to 580	1000	750	625	500	438	375	313	375	250	231	213	194
581 to 600	1020	765	638	510	446	383	319	281	255	236	217	198
601 to 620	1040	780	650	520	455	390	325	286	260	241	221	202
621 to 640	1060	795	663	530	464	398	331	292	265	245	225	205
641 to 660	1080	810	675	540	473	405	338	297	270	250	230	209
661 to 680	1100	825	688	550	481	413	344	303	275	254	234	213
681 to 700	1120	840	700	560	490	420	350	308	280	259	238	217

Source: Reprinted from Ronald H. Ballou, *Basic Business Logistics* (Englewod Cliffs, N.J.: Prentice-Hall, Inc., 1978), Table 7–4, p. 174.

The freight bill for the shipment of clotheslines (treating it as a carload shipment of 24,000 pounds) will be

$$\$1.27 \times 240 = \$304.80$$

Paying for that extra 2,000 pounds of "phantom" freight results in a saving of $126.40! Most certainly, paying for the additional pounds and qualifying the shipment as a carload makes good sense—and is also quite legal.

TRAFFIC MANAGEMENT

What does the traffic manager of an organization do?

> TRAFFIC MANAGEMENT: The control and management of transportation services.

The functions performed by the traffic manager are numerous and include the following:

- Rate determination: Finding the lowest legal rate for the shipment of company products.
- Rate negotiation: Obtaining new and usually lower rates for company products.
- Commodity descriptions: Since the description of a commodity has an important bearing on the classification, proper descriptions are critical. Brand names are not enough.
- Selection of carriers.
- Preparing documentation.
- Routing where permitted by the carrier.
- Filing loss and damage claims.
- Filing reparations claims.
- Avoiding the payment of demurrage and detention charges.
- Freight bill audits.
- Shipping and marking hazardous materials.
- Consolidating small shipments.
- Operating and managing company-owned transportation equipment.

Several items in this list are fairly technical in nature and require more explanation to be fully understood by logistics managers.

To mention the importance of carrier selection gives the initial impression of belaboring the obvious. What is not so obvious is the extent to which carriers even in the same mode of transportation differ from each other. They differ in the rates charged, dependability of service, speed of service, loss and damage record, and special services provided. The list of these services can also be lengthy, among which are:

- Diversion, or changing the destination of a shipment while still en route.
- Reconsignment, or rerouting a shipment to a new destination after it has arrived at the original destination.
- Transit privileges, such as stopping the shipment en route to partially load or unload the freight.
- Protective services, such as refrigeration for fresh fruit, vegetables, and other perishables, or heat to prevent damage due to cold.
- Tracing delayed shipments.
- Expediting, or rushing a car through the marshalling yards.
- Switching cars between the main line and a private siding.

Not all companies need these special services, but quite a variety is available from the carriers. A chemical manufacturing firm, for example, might need a heated car to prevent the substances it is shipping from being damaged by low temperatures.

The preparation of freight documents is of particular importance and must be done with great attention to accuracy. Two documents deserve mention here: the freight bill and the bill of lading.

The *freight bill* contains an item-by-item description of the contents of the shipment and the freight charges. The *bill of lading* also contains a detailed description of the freight, but omits the freight charges. Bills of lading perform three basic functions:

- A receipt to the shipper certifying that the carrier has received the items.
- A contract for the service of transportation between the shipper and the carrier.
- A certificate of title.

Not all bills of lading perform all three of these functions at one time. What the bill of lading does depends on the type of bill:

- Straight bill of lading: Requires that the shipment be delivered to the consignee named on the bill.
- Order bill of lading: Buyer or consignee must pay for the shipment and claim the bill (frequently at a bank) to receive the shipment. This form functions as a certificate of title to the goods: Whoever has the bill owns the merchandise.
- Long form: Contains the complete transportation contract in microscopically fine print on the back of the form.
- Short form: Simply refers to the "Uniform Straight Bill of Lading," without containing any of the text of the transportation contract.
- Preprinted form: A bill of lading of any of the preceding types on which the shipper has printed entries on products frequently shipped.

Accuracy in describing the products being shipped on the bill of lading is critical since the proper description of a commodity is a major determinant influencing the freight classification into which the item is placed. We know from a previous discussion the effect classification has on the ultimate money rate paid for the shipment.

Filing loss and damage claims is a regrettable though frequent necessity in transportation. Shipments get lost and goods get damaged in transit for any number of reasons. The law requires that the claim be settled between the shipper and the carrier within 120 days of filing the claim. The amount of damages claimed, however, must equal the actual cost. The cost to be claimed is determined by the end use of the product.

The courts (unresolved disputes concerning loss and damage claims go to the courts, not the Interstate Commerce Commission) have decided that wholesale cost is the proper cost when the items are going into inventory; retail cost is the relevant cost when the items are going to a customer.

Reparations are due the shipper when someone discovers that the carrier has charged an illegally high rate. Note the importance here of freight bill audits. Whether or not reparations can be claimed hinges on the difference between a legal and a lawful rate. Any rate published in a tariff is a *legal* rate. Occasionally, a published rate can still violate existing transportation law. For example, the carrier might be charging a higher rate for a shorter distance when included in a longer haul, thus violating the long-haul, short-haul clause of the Interstate Commerce Act. A rate is a *lawful* rate when it is both published in a tariff and is not in violation of any law. A legal rate can thus be either lawful or unlawful. The unlawful rate is the one that furnishes the basis for a reparations claim.

The traffic manager can save the company sizable sums by avoiding the payment of demurrage charges (in rail transportation) or detention charges (in highway transportation). They both involve the same poor transportation management practice: holding carrier equipment beyond the free period and, in effect, using such equipment as mobile warehouses. *Demurrage/detention* is thus a penalty imposed on the shipper for delaying the loading or unloading of equipment beyond the usual 48-hour free period (which customarily does not include weekends and holidays). The auditing of freight bills can detect not only over- and undercharges, but the payment of excessive penalty charges as well.

The handling of hazardous materials is a matter that extends far beyond the warehouse of the company that ships such materials. Such products are extremely dangerous to human life, equipment, and the environment. Once in a while the TV screen is filled with a picture of a great plume of black smoke billowing from a flaming tank car derailed in the middle of a town. Shipments of radioactive materials leaking radiation have been known to expose every roll of film in every photography store along the route traveled. In fact, we are plagued by the question of the safe disposal of waste chemicals and other poisonous substances.

As a result of the perils involved in shipping this type of product, containers of the requisite strength with the special markings prescribed by law and regulations must be used. To ensure compliance with these laws, carriers routinely specify the containers to be used and the marking placards (more will be said about these in Chapter 7) required in their tariffs. Explosives, poisons, flammables, and radioactive materials all fall under these special rules. Even the method of loading the containers aboard carrier equipment and the type of blocking and bracing to be used are prescribed in detail. In other words, the law is very specific on how to pack, ship, and mark hazardous freight, and for good reason. The loss of life, property, and harm to the environment can be considerable should these regulations be evaded by unwise and careless shippers.

Consolidating small shipments involves stopping the shipment in transit to break a large shipment into smaller ones or to combine a number of small shipments into a single larger one. Break bulk involves sending the products in a single large shipment part of the way to take advantage of carload rates to a central point. Here the one large shipment is divided into less carload (and more expensive) shipments for the rest of the journey to the separate customers or destinations. For example, Figure 5-1a illustrates a situation that might exist before a distribu-

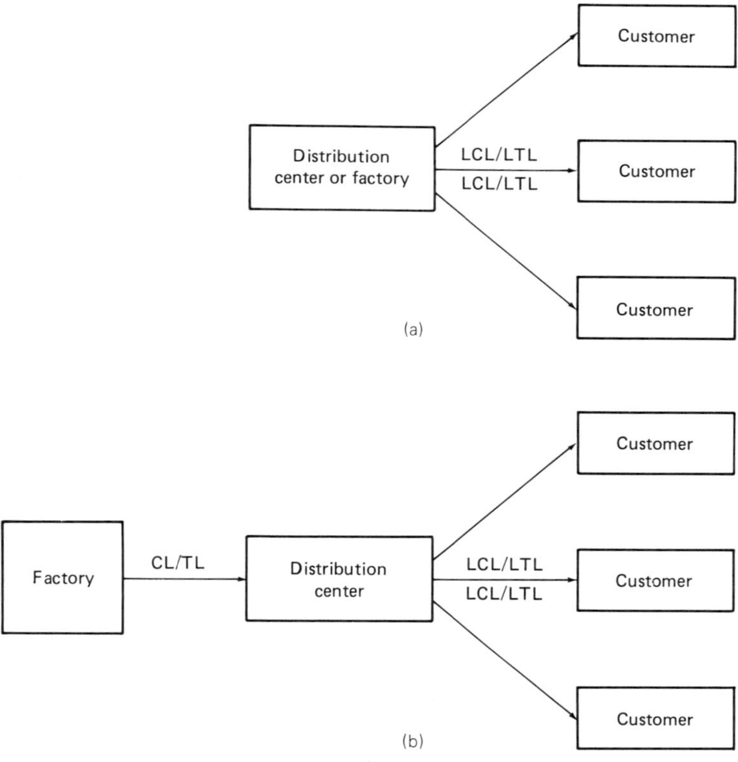

FIGURE 5-1
(a) Shipment direct from DC or factory to multiple destinations; (b) shipment direct from factory to a break-bulk DC

tion center was used to break bulk. Figure 5-1b shows the same situation but after a break-bulk point is in use.

In contrast, make bulk involves shipping small less carload shipments for as short a distance as possible to minimize the less carload freight charges from a branch warehouse to a distribution center. Here the small-sized shipments are consolidated into carload lots and then sent on the rest of the way to the customer. For example, Figure 5-2a shows a situation that might exist before a distribution center is used. The situation after a make bulk is in use is shown in Figure 5-2a. Considerable cost savings on transportation charges can be realized by shipping the carload lots as far as possible and minimizing the distance shipped in less carload lots.

The last function of the traffic department to be discussed here

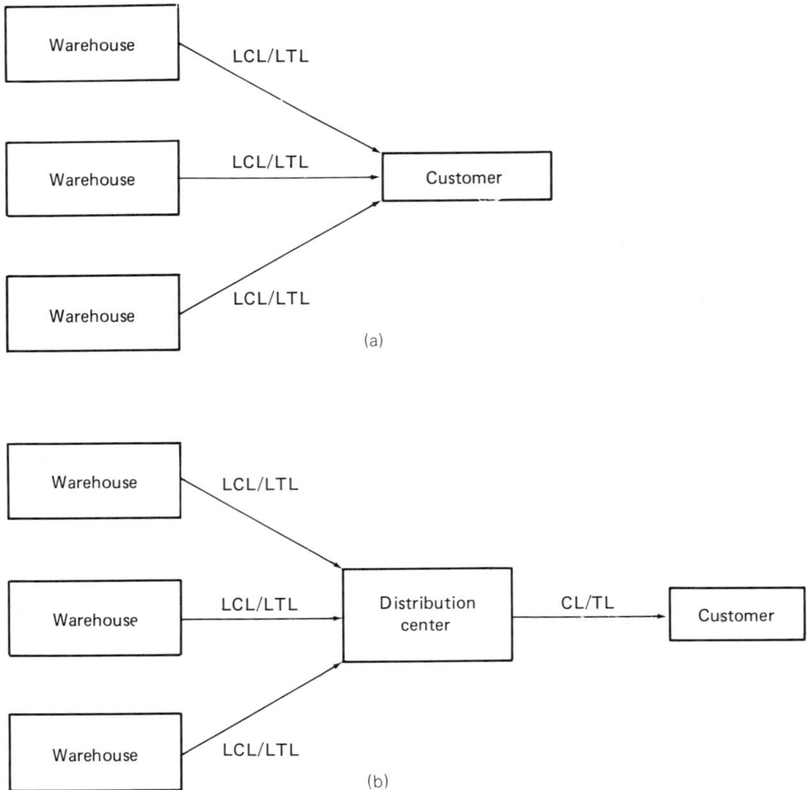

FIGURE 5-2
(a) Shipment from multiple storage locations to customer; (b) shipment from warehouse to a make-bulk DC

gives us the opportunity to demonstrate how a quantitative managerial technique can be used to improve the quality of management: the management and operation of a company-owned means of transportation.

Many firms use their own equipment to deliver shipments to their customers or carrier freight stations rather than use a commercial for-hire carrier. In such cases, the problem must be faced of routing the transportation equipment in the most advantageous manner. If management wishes to route a delivery truck, for example, to minimize the distance or total time required to deliver a number of shipments to customers, a process called *heuristic programming* can be used. In this example, "heuristic" means a process of trial and error, but on an organized, scientific basis.

Suppose the traffic manager is asked to design an optimal routing (i.e., a routing requiring the least distance to be traveled) between the locations of five different customers. The distance (or driving time if a unit of time is being used) between each of the five points is best arrayed in a matrix, much like the tables that show the distances between cities found in many books of road maps. The matrix in this instance would be like Table 5-4, with the points to be served along the top and right margin and the numerals showing distance. This arrangement will permit us to consider every point in turn until we have the best routing from our plant to each of the customers, and then return to the plant.

To start the programming process, consider first a round trip from the warehouse at point A to the first customer location at point B. The distance between each point is indicated by circled numerals as follows:

A (15) B (15) A = 30 miles for a round trip

The second line of the matrix is formed by adding the location of the

TABLE 5-4
Customer Locations and Miles between Each Customer

A	*B*	*C*	*D*	*E*	
	15	23	32	17	A
		20	27	18	B
			25	19	C
				20	D
					E

A through E indicate customer locations.

second customer at point C and inserting it and the miles involved between points A and B, as follows:

A (23) C (20) B (15) A = 58 miles

Now form the next line of the matrix by inserting point C between points B and A:

A (15) B (20) C (23) A = 58 miles

These last two lines tell us that whether we serve point B before point C, or vice versa, the distance factor will be the same. As a matter of standard procedure, we always select the first routing when other following routings yield the same distance. In other words, for our next major step we use the routing A–C–B–A (even though the second routing, A–B–C–A requires the same travel distance).

Now add point D to the routing just selected by first inserting point D between points A and C, as follows:

A (32) D (25) C (20) B (15) A = 92 miles

Note how the insertion of point D displaces all the other points one place to the right. Inserting point D between points C and B changes the distance factor as follows:

A (23) C (25) D (27) B (15) A = 90 miles

We complete the testing of insertions of point D by placing point D between points B and A to see what results:

A (23) C (20) B (27) D (32) A = 102 miles

At this point, a pause is in order to recapitulate what has been done thus far. The insertion of point B in the first step and successive insertions of points C and D appear as follows: by adding B,

A (15) B (15) A = 30 miles

By adding C,

A (23) C (25) B (15) A = 58 miles
A (15) B (25) C (23) A = 58 miles

For procedural consistency, we arbitrarily selected the first of the two lines generated by adding point C since the routings required the same distance of 58 miles for the round trip. Using the line A–C–B–A, we then inserted point D on successive lines between each of these points. By adding D,

A (32) D (25) C (20) B (15) A = 92 miles
A (23) C (25) D (27) B (15) A = 90 miles
A (23) C (20) B (27) D (32) A = 102 miles

Since we are trying to minimize the distance traveled by our delivery truck, we would select the second row, indicating a routing that requires 90 miles to be traveled, for use when inserting point E. By adding E,

A (17) E (19) C (25) D (27) B (15) A = 103 miles
A (23) C (19) E (20) D (27) B (15) A = 104 miles
A (23) C (25) D (20) E (18) B (15) A = 101 miles
A (23) C (25) D (27) B (18) E (17) A = 110 miles

As a check on the accuracy of the programming, the last point used (in this case point E) will appear in the final matrix on a diagonal line from upper left to lower right. Since we are trying the successive point insertions to locate the shortest travel distance, we would select the routing that requires 101 miles along the points A–C–D–E–B–A for the round trip.

Companies with a number of delivery vehicles that travel to several locations each day find this technique especially useful by letting the computer construct the successive matrices of points to locate the optimum routing. Keep in mind, however, that factors other than just distance or time might require human intervention in this problem-solving process. One or two customers may deserve priority treatment; thus the order in which the deliveries are to be made would be changed by such "other considerations." The routing actually used would then be less than the optimum indicated by the procedure. This is why quantitative techniques cannot always solve problems and why managers draw salaries. Such techniques can be valuable in assisting management, but human factors can always intervene, leading to alternate solutions that are less than optimal mathematically but still equally desirable from a human point of view. If the other factors are important enough, good human judgment justifiably will overrule the computer.

OBJECTIVES ACHIEVEMENT CHECKUP

The following exercises will help you in assessing the extent to which you have achieved the objectives for this chapter. Where essay-type answere are required, the more you are able to condense your knowledge into a short paragraph of a few sentences the better. Looking back as you work these exercises and rereading parts of the chapter are permissible; but do not look ahead to the answers until you have done your best with the questions.

1. How was the idea of public regulation of private enterprise introduced into the United States?
2. What is a carrier legally obligated to do?
3. List a few of the different types of rate structures.
4. A company has 26,000 pounds of insulating tape packed in boxes to ship from Benton Harbor, Michigan, to Athens, Ohio. Calculate the lowest freight charge.
5. Explain the difference between a straight bill of lading and an order bill of lading.

ANSWERS TO CHECKUP QUESTIONS

1. The case *Munn* v. *Illinois (1877)* in which the Supreme Court of the United States permitted the state of Illinois to set rates charged by grain elevators.
2. A carrier is legally obligated to:
   a. Serve all comers within the terms of its legal classification (e.g., common, contract, private, or exempt carrier).
   b. Accept small shipments even if the revenue does not cover the cost of providing the service.
   c. Deliver freight safely.
   d. Assume liability for loss or damage.
   e. Charge reasonable rates and not discriminate.
   f. Observe federal, state, and local laws governing speed limits, weight limits, and safety.
3. A few of the types of rate structures are:
   a. Volume: The quantity of freight hauled heavily influences the rate (e.g., less carload rates are higher than carload rates for a particular product).
   b. Distance: Rate increases as distance hauled increases.
   c. Differential: Rate is determined by adding or subtracting a specific amount (i.e., the differential) from a standard rate.

4. First gather the available information from the freight classification and rate tariffs:

   LCL rating: $77\frac{1}{2}$

   Minimum CL weight: 30,000 pounds

   CL rating: 45

   Rate basis number: 351

   At first it looks as though this shipment weighing 26,000 pounds will not meet the minimum carload weight of 30,000 pounds and thus will not qualify for the carload rating. Using the LCL rating of $77\frac{1}{2}$, the rate would be

   $$\$1.52 \times 260 = \$395.20$$

   with a rate basis number of 351.

   Now calculate the freight charge if the company is willing to pay for 4,000 pounds of "air" and call the shipment a carload. Unfortunately, the rate tariff does not contain a column for class 45, the carload rating. As a result, we must first find the rate applicable to class 45:

   $$\text{Class 100 rate} \times 45\% = \text{Class 45 rate}$$

   $$\$1.96 \times 45\% = \$0.882$$

   Now calculate the freight charge as though the shipment were a full carload of 30,000 pounds:

   $$\$0.882 \times 300 = \$264.60$$

   Paying for the extra 4,000 pounds and qualifying for the carload rate yields a significant saving:

LCL freight charge	\$395.20
CL freight charge	264.60
Saving	\$130.60

5. a. The straight bill of lading requires that the shipment be delivered to the consignee named on the bill.

   b. The order bill of lading requires the consignee to obtain the bill of lading by paying for the shipment before claiming the freight.

part 2

OPERATING THE LOGISTICAL SYSTEM

6

Warehousing: Functions and Location

CHAPTER OBJECTIVES

Upon completion of this chapter, you should be able to:

A. Appreciate the importance of siting in the operation of a firm.
B. List the ways of acquiring storage space.
C. Compute locations by the center-of-gravity method.
D. Use the distance–cost method of siting to determine location when considering production costs.
E. Understand warehouse functions.

Most individuals entering upon a career in logistics join organizations whose facilities are already sited. Nothing, however, remains perfectly stable in commerce and manufacturing. This constant change occasionally creates the need either to relocate an existing plant or warehouse or to find a location for an entirely new operation. Deciding where a facility will be located is a decision of critical importance both financially in the immediate present and for the future survival of the company.

SITING FACTORS

Those who spend their lives in the real estate business have three major rules for success: location, location, location. The logistician can take a page from that book with great benefit. But how to decide on a suitable location?

First, what type of facility are we talking about: a factory, a central warehouse to store goods, or a distribution center? If a distribution center, then what is the major purpose: to match supply with demand for the product or to assist in marketing by improving customer service? The most probable answer is a blend of matching and marketing.

Second, whether the facility is a manufacturing plant, a storage facility, or a distribution center, certain factors must be taken into consideration.

Since factories and warehouses, in common with customers, need their own logistical support, attention should be paid to the availability of resources. If the site includes a manufacturing plant, access must be had to raw materials, components, and other materials used in the manufacturing process. Distribution centers may not need sources of raw materials, but they do need labor. The ability of the surrounding region to provide a source of workers for factory and warehouse alike is another important consideration.

Rather than fall into the trap of regarding labor as of uniform quality, the available work force needs to be assessed from a qualitative as well as a quantitative point of view. Manufacturing will require a mix of skills that only a personnel manager could appreciate in its sheer complexity. Distribution centers also require an array of skills that must either exist at the time of hiring or that can be acquired through formal and on-the-job training programs. The individuals charged with the acquisition of property should also understand the need for the availability of suitable land, including use restrictions or zoning that would limit how the land is to be used. And while speaking of location, give a thought to climate.

Something must be attractive about California and the Sunbelt of the Southwest to business. Why endure the climatic misery of ice and snow when relief is available in Florida and points west? At the same time, something can also be said for other sections of the nation that do not include visions of lolling about on sun-drenched beaches. Other factors, perhaps of equal importance to climate, are a skilled local work force and a plentiful supply of excellent educational facilities to replenish the supply. Each regional area has its advantages and disabilities, ranging from the attractive winter recreational facilities in New England to the dense transportation networks of the Midwest and the accessibility to markets that these modes of transportation provide.

Acquiring Storage Space

Storage space can be acquired in at least four different ways: own, by building an entirely new structure or buying an existing structure; rent on a short-term basis; lease on a long-term basis; or store in transit, using the equipment of the carrier as a sort of rolling warehouse.

Should a company decide to own a plant or distribution center, the business will gain the advantage of having complete control. Despite the cost of construction or purchase, this could turn out to be the least expensive method if the use of the facility is extensive throughout the year. From an operational point of view, the building can be tailored to the precise needs of the company. From a financial point of view, the company would also get the advantage of a deduction for depreciation from the corporate income tax on the building and equipment and further deductions for such allied expenses as utility costs, maintenance, and taxes paid to local jurisdictions.

Should the needs of the company be fairly short term (e.g., less than a year), renting may be the best alternative to owning. If the need for space varies considerably from time to time, additional square footage can be added to the rental agreement even if the space is required for periods as short as one month. A distinct advantage to a cost-conscious company is the absence of any requirements to invest in equipment, since most commercial warehouse operators provide materials-handling equipment. Rental space can also be varied as needs develop for the storage of liquids, products requiring cold storage, or general-purpose space.

A longer-term commitment (e.g., in excess of a year) with all the advantages of renting can be obtained with the lease. If necessary, the lease could cover the entire building. The leased facility could be either a warehouse owned by private individuals or offered for lease by some public agency (e.g., a city or county government).

Storage in transit saves the company the trouble of operating a physical facility. This method uses transportation equipment as a moving warehouse. If the storage time needed by the company is short enough to coincide with the time a carrier takes to haul goods from origin to destination, then the rolling stock of the carrier can be used as storage. If the customer is temporarily unable to receive a shipment, delivery can be delayed by using a slower mode of transportation. This method, as you might suspect, requires a particularly efficient logistical system that can schedule transportation pickup and delivery with great precision so that products are made available for shipment exactly when the equipment is ready to haul them, and delivery can be made to the customer neither too late nor unacceptably early.

Selecting a Location

Site selection should follow a logical sequential process: First, the region is selected (e.g., New England, Pacific Coast, Midwest, or the Sunbelt); then follows selection of the state, county, and city; finally, the exact parcel of land is chosen. Let's take a close look at some of these factors in greater detail.

The nature of the product will determine whether the facility will be located close to the source of raw materials or farther from the raw materials and closer to the market center where the products are sold. A commodity such as logs, which lose considerable weight while being transformed into boards, would require a site close to the forest, since transportation charges on the heavier logs will be more than on the lighter, finished lumber. Such a situation is illustrated in Figure 6-1a.

A product such as gravel, which gains weight in the process of being transformed into cement, would indicate a site closer to the market center, since transportation charges on the heavier bags of cement will be higher than on the bulk gravel. This situation is illustrated in Figure 6-1b.

The legal aspects of labor present a reality that must also be faced. Many employers would prefer a situation in which labor unions were either weak or totally absent. However we might feel personally about labor unions, the strength of unions in the projected area of location could influence the decision. The prevailing legal climate concerning labor would be an additional consideration. For example, in states that have right-to-work laws, the closed shop (i.e., no one will be hired unless they already belong to the union) is illegal.

Taxes are no small matter of concern, especially since some taxing jurisdictions provide tax breaks for a period of years to businesses that agree to locate in their areas. The amount of taxes levied on a business must be balanced against both what is to be taxed (e.g., a

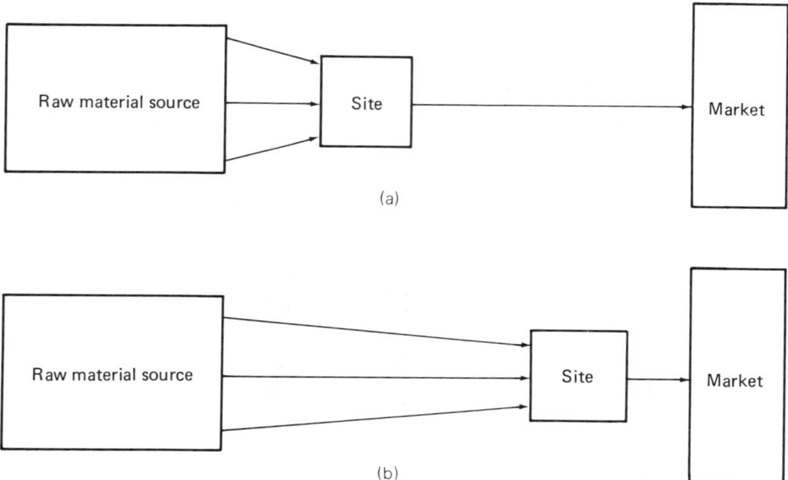

FIGURE 6-1
(a) Location of factory site processing raw materials that lose bulk; (b) location of factory site processing raw materials that gain bulk

favorite and often burdensome tax is levied on inventories) and what the taxes buy: water and sewage services, police and fire protection, and, for the employees, schools and housing.

Energy costs are of particular importance since they tend to be rising in almost every locality. Electricity costs depend in good measure on how the power is generated. Is the power supplied cheaply by the Tennessee Valley Authority from the water power of its dams? Is the electricity generated by some high-cost fuel such as gas or oil or is low-cost coal used? Or is the split atom heating the steam that turns the generating turbine?

The customer looms large in any planning for site selection. Frequently, a facility needs to be near customers who have buying power for the product sold. Especially in the case of manufactured goods, location near the larger customers would be sensible. In this way, the type of customer influences site selection. If the types of goods produced are primarily for the general public, the facility should be capable of being moved on occasions to follow shifts in population. If, in contrast, the goods are sold to industrial consumers, the facility should be located where costs are lowest to keep the price of the product as low as possible and still generate a satisfactory product. Industrial goods also tend to be produced in factories that cannot be moved easily. All right, so a low-cost location is desirable. How does one go about finding this operational Utopia?

FINDING THE LOW-COST LOCATION

The *center-of-gravity method* can be used to find the average distance between the storage location and the retail outlets of the customer or the company itself.

Initially, let us consider only the distances involved. This requires setting up a grid resembling the coordinate squares on a map, but using only east and north distances on the grid, as illustrated in Figure 6-2. Points on the grid are plotted according to the map-reading rule: "Read right, then up." This means the east coordinate is read first along the *X* or east axis of the grid; then the north coordinate is read up from that point along the *Y* or north axis of the grid. Figure 6-3 illustrates the manner of identifying each point located in this way: The number of miles to the east of the zero origin of the grid is entered in the upper left quadrant, the number of miles to the north is entered in the upper right quadrant, and the name or designation of the point is written across the bottom. Thus, site D, which is 13 miles to the east and 8 miles to the north, would be marked as shown on the right side of Figure 6-3.

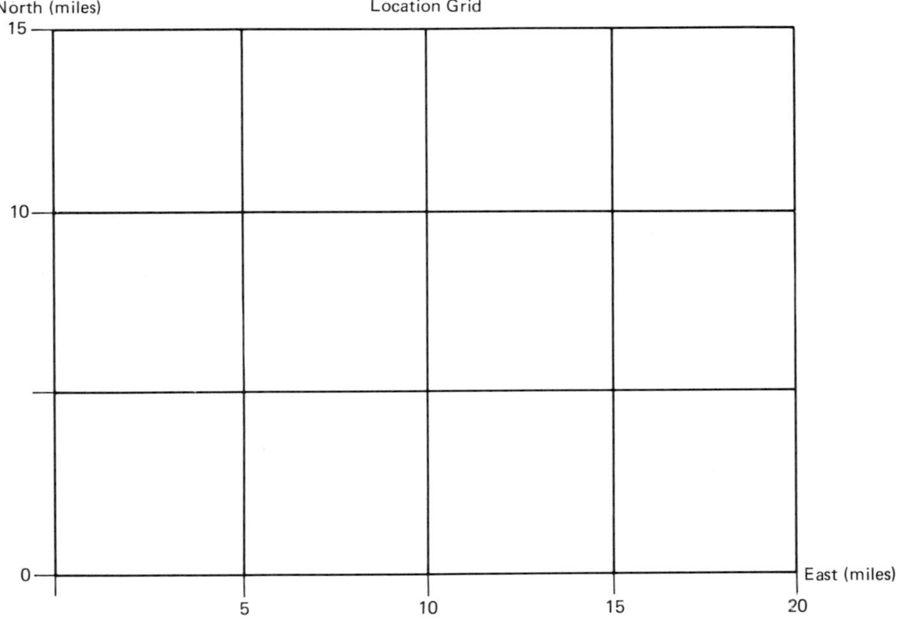

FIGURE 6-2
Location grid

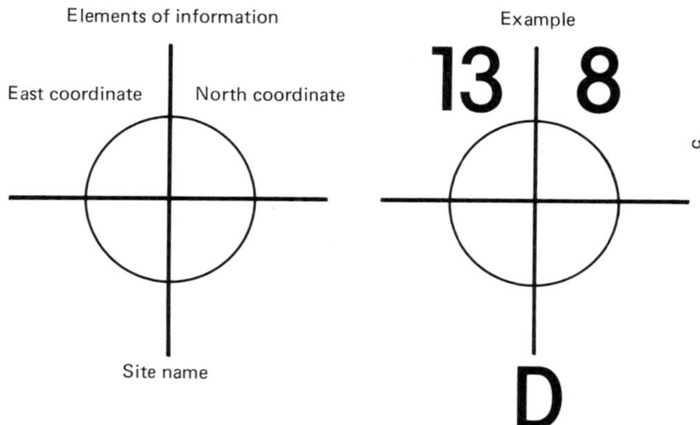

FIGURE 6-3
Site location marker

Suppose we have eight sites to plot, labeled from A to H, which we will use to find the best location for a distribution center serving the eight customers:

Customer Location	*Coordinates*	
	East	*North*
A	2	9
B	5	8
C	3	3
D	13	8
E	15	4
F	18	3
G	11	2
H	6	4

Our next task is to plot these eight locations on the grid, as in Figure 6-4. To find the optimum location of the storage site (the site at the center of gravity of the customer locations), we need only compute the arithmetic averages of the east coordinates and north coordinates and plot these two averages on the grid. In this case, the average of the east coordinates is $9\frac{1}{8}$, while the average of the north coordinates is $5\frac{1}{8}$. When we plot this center-of-gravity location at $9\frac{1}{8}$ east and $5\frac{1}{8}$ north, we have a neat, maplike display of the desired storage location and the customers it will serve. We have starred the desired location for our

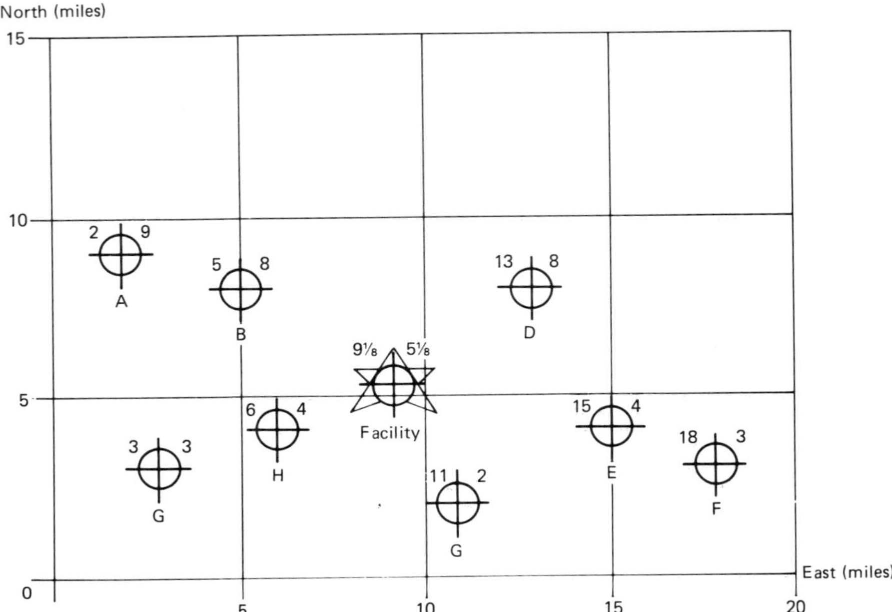

FIGURE 6-4
Plotted site location

facility on Figure 6-4 to differentiate it from the customer locations that are designated by circles.

In most cases, distance is only one factor to be considered. A combination of distance and transportation costs frequently determines which markets a firm will be able to serve. The center-of-gravity method described provides a rough estimate of where the site should be. A more precise way modifies the center-of-gravity method by combining both distance and transportation costs. This is done by using the east and north grid coordinates as weights in computing a weighted average.

Suppose a company has a factory at point A that normally produces 80,000 units of a product. The factory distributes these 80,000 units directly to four customers. The amounts of this product usually shipped to each customer, the unit and total transportation costs, and the east and north grid locations are given. To relieve the factory of the shipping function, the company is considering locating a storage and distribution facility in the area. Data have been gathered to assist in determining the optimum location of the new facility, as follows:

Customer Location	*Quantity Shipped*	*Unit Transportation Cost*	*Total Cost*	*Coordinates*	
				East	*North*
Factory A	80,000	$0.15	$12,000	2	3
Customer B	15,000	0.35	5,250	1	1
Customer C	30,000	0.20	6,000	3	5
Customer D	25,000	0.25	6,250	4	2
Customer E	10,000	0.40	4,000	5	4

These locations are then plotted on a grid as illustrated in Figure 6-5. Given this information, where would be the best location for a warehouse or distribution center?

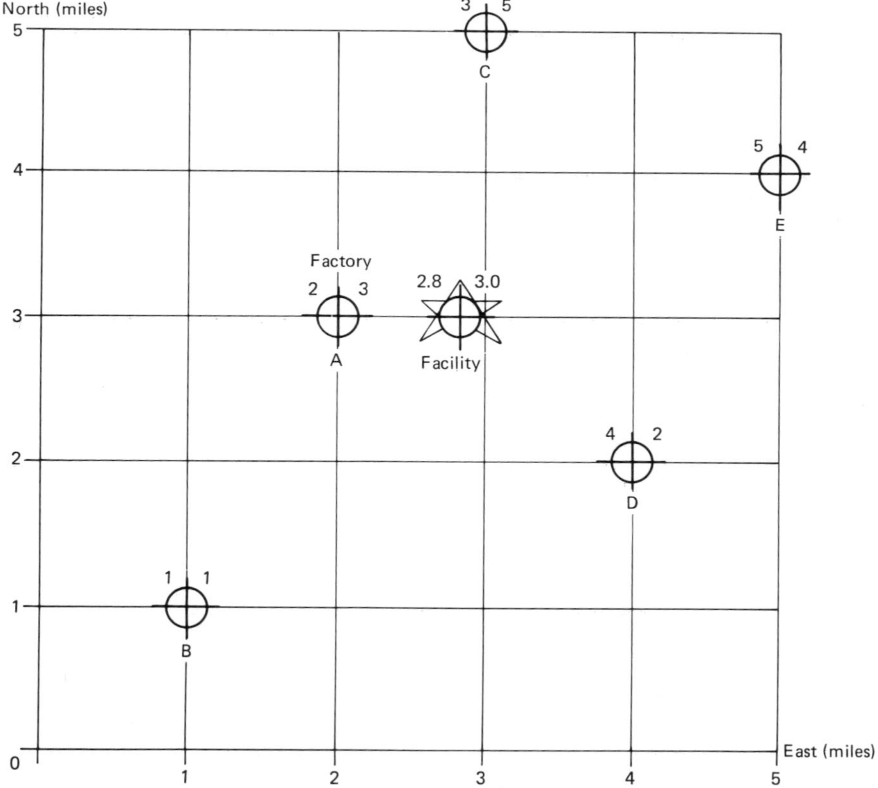

FIGURE 6-5
Plotted site location

The optimum storage location would be computed by using the grid coordinates and transportation costs to find a weighted factor for each location on the grid. A location factor for each east coordinate is calculated using the following formula:

East location factor = total transportation cost × east coordinate

And a second factor for each north coordinate is computed in the same way using the following formula:

North location factor = total transportation cost × north coordinate

The computations for these factors would appear as follows:

Customer Location	*Transportation Cost*	*East Coordinate*	*Factor (cost × E)*	*North Coordinate*	*Factor (cost × N)*
Factory A	$12,000	2	$24,000	3	$36,000
Customer B	5,250	1	5,250	1	5,250
Customer C	6,000	3	18,000	5	30,000
Customer D	6,250	4	25,000	2	12,500
Customer E	4,000	5	20,000	4	16,000
Totals	$33,500		$92,250		$99,750

The grid location of the new facility being planned is now computed using the totals indicated for the transportation costs and both east and north factors as follows:

$$\text{East coordinate} = \frac{\text{total E factors}}{\text{transportation costs}} = \frac{92{,}250}{33{,}500} = 2.8$$

$$\text{North coordinate} = \frac{\text{total N factors}}{\text{transportation costs}} = \frac{99{,}750}{33{,}500} = 3.0$$

These grid coordinates of 2.8 east and 3.0 north are plotted on the grid in Figure 6-5 and marked with a star.

Either of the center-of-gravity methods illustrated provides a reasonable first approximation to the optimum location of the planned facility. The final location would be influenced as well by such factors as the availability of land, the cost of building or leasing the facility, and other factors mentioned earlier in this chapter.

One element not included in the data of the center-of-gravity method is the cost of production. Common sense tells us a buyer can be expected to give an order to the lowest-cost supplier. This means that the supplier with the lowest-cost combination of production and transportation costs will get the order. In effect, transportation costs in the *distance–cost method* determine how far from the factory the company will have a cost advantage over competitors who have the same production costs but higher transportation costs.

This method requires some simple elementary school algebra in finding the value of one unknown quantity in the model equation. The distance equation is

$$D = C_1 + C_2$$

where D = distance between two competitors

C_1 = distance to the boundary of the market for competitor 1

C_2 = distance to the boundary of the market for competitor 2

The equation for finding the boundary between the two competitors and their market areas at which costs are equalized is

$$PC_1 + TC_1 C_1 = PC_2 + TC_2 C_2$$

where PC_1 = production costs for competitor 1

PC_2 = production costs for competitor 2

TC_1 = transportation costs for competitor 1

TC_2 = transportation costs for competitor 2

C_1 = distance to the boundary of the market for competitor 1

C_2 = distance to the boundary of the market for competitor 2

We will be able to locate the boundary of the market area for each of these two competitors if we have the following information:

- Distance between the two competitors is 200 miles
- PC_1 = \$10 per unit of output
- TC_1 = \$0.25
- PC_2 = \$15 per unit of output
- TC_2 = \$0.15

Let's use the distance equation first. If the two companies are 200 miles apart as given, then

$$D = C_1 + C_2 = 200 \text{ miles}$$

and, by switching around the terms of this equation a bit,

$$C_2 = 200 - C_1$$

Notice what we have done by this small manipulation of terms in the distance equation. Without using any other information thus far, we have derived a value for C_2, which is one element in the boundary location equation. Now substitute in the boundary equation what we know:

$$PC_1 + TC_1 C_1 = PC_2 + TC_2 C_2$$

We know that

$$C_2 = 200 - C_1$$

And we know the production and transportation costs, which, when substituted, give us a boundary equation looking like this:

$$\$10 + 0.25C_1 = \$15 + 0.15(200 - C_1)$$

Removing the parentheses leaves us with the equation looking like this:

$$\$10 + 0.25C_1 = \$15 + 30 - 0.15C_1$$

but with only one unknown quantity, the C_1.

Placing the C_1 by itself to the left of the equals sign gives us the same equation we have been working with, but in this form:

$$0.25C_1 + 0.15C_1 = \$15 - \$10 + 30$$

Now ignore the dollar signs and just use the numbers:

$$0.40C_1 = 35$$

$$C_1 = 87.5 \text{ miles}$$

This result means competitor 1 should locate 87.5 miles from the market boundary. To find out how far from the boundary competitor 2 should locate, substitute the value for C_1 we have just calculated into the original distance equation:

$$C_2 = 200 \text{ miles} - C_1$$

$$C_2 = 200 \text{ miles} - 87.5 \text{ miles}$$

$$= 112.5 \text{ miles}$$

For purposes of location, these calculations indicate the following: Given the distance of 200 miles between the two competitors, competitor 1 has the cost advantage from that location to within 112.5 miles of competitor 2. Thus competitor 1 is 87.5 miles from the border marking the line of equal costs. If competitor 1 tries to go beyond that 87.5 miles and penetrate into the 112.5 miles within which competitor 2 has the cost advantage, competitor 1 is going to need something more than just price to attract customers. Better and more reliable service would be two good points in favor of competitor 1.

Other location models are available, but they have the disadvantage of requiring computer assistance. The center-of-gravity and distance–cost models illustrated here have the advantage of being both simple and workable manually, while giving a decent approximation of an optimum solution.

Although not a mathematical model, the *free-trade zone* is worthy of consideration in the site-selection process. The free-trade zone is an area set aside by a national, state, or local government in which the payment of import/export duties or other taxes is suspended until a shipment leaves the zone. Merchandise can be shipped into a free-trade zone without the payment of any taxes and then stored there until shipped out. While in the zone, the products can be exhibited to prospective customers and even sold on the spot. Where the facilities to do so are available, some light manufacturing or assembly work can be done to make the product more salable. Only when the products finally leave the free-trade zone will taxes be imposed. In this way, the payment of taxes is deferred until the last possible moment, to the advantage of the shipper or other party who must pay the taxes or duties.

Relocating a Facility

Population shifts, acquisition of new customers at some distance from the usual place of business, revisions in transportation charges, the appearance (or disappearance) of modes of transportation—all may require finding a new location for a manufacturing or storage facility. When the decision is made that a change is needed, several alternatives present themselves: expand the existing site or, if that will be satisfactory, move to a new site on which a facility either already exists or can be built.

An important consideration in moving to a new location is employee morale. A company that decides to move a considerable distance from the old site had better count on losing some employees. People resist the radical change in their lives that relocations cause and would prefer to remain where their friends and families comprise familiar and loved surroundings, even if this means finding employment with some other firm. Even some equipment may not make the move. This will involve decisions concerning what equipment can be taken to and used at the new site and what items of machinery should be scrapped or sold.

WAREHOUSE OPERATION

Up to this point, the words "warehouse" and "distribution center" have been used as though they were synonyms. This is not strictly true, so let us now distinguish clearly between the two types of facility:

> WAREHOUSE: A logistical facility whose primary purpose is the storage of goods.
>
> DISTRIBUTION CENTER: A logistical facility whose primary purpose is the movement of goods.

A clear distinction between these two installations should be made. Goods move into a warehouse and sit there until someone gives the order to move them out. The distribution center seeks to keep the goods moving as rapidly as possible. The distribution center can be production oriented by holding stockpiles of goods until needed by the factory; or the center can be market oriented by holding finished goods until dispatched to a customer. The main difference is one of emphasis: passive storage in the warehouse, but active movement in the distribution center. Both, of course, maintain goods on hand until wanted.

What Storage Facilities Do

Storage facilities are on the receiving end of orders. These orders can be from the factory for raw materials, assemblies, or other components, or the orders can arrive from customers. An important function of the warehouse or distribution center is consolidation of shipments to take advantage of lower transportation costs on the larger-sized shipments.

Goods arrive and leave in one of two forms. Goods that arrive in carload or truckload quantities usually need to be broken down into smaller shipments to individual customers (e.g., the break-bulk function mentioned earlier). Goods that arrive in less carload or less truckload lots usually must be assembled into the larger carload or truckload lots (e.g., the make-bulk function, also mentioned earlier). A balancing act is involved here: The savings to be gained from large-volume shipments must be balanced against the cost of holding goods until the volume quantity is accumulated.

Types of Storage Facilities

The *public warehouse* is similar to a common carrier. The warehouse operator must provide use of the facility to all who wish, provided the facility is equipped to handle the merchandise to be stored. A warehouse, for example, that can handle only general merchandise could not be compelled to store bulk liquids. Public warehouses are of two varieties: a general-purpose facility and a special-purpose facility.

In a general-purpose warehouse, only the space actually required need be rented by the company. No capital investment is required in either building or materials-handling equipment. This type of warehouse tends to be highly flexible in the amount of space offered. The services offered by the general-purpose public warehouse include the following:

- Bonded storage, which means fees, duties, or taxes need not be paid until the shipment leaves the warehouse.
- Assembles outbound shipments for the shipper.
- Maintains a desired level of inventory for the shipper.
- Provides data-processing equipment compatible with the machines used by the customer of the warehouse (and if the warehouse is chosen carefully, it might even be compatible with the machines used by the customers of the shipper).
- Provides physical security for goods being used as collateral for a loan (e.g., the warehouse receipt can be presented as security for the loan since the warehouse operator will not permit the removal of any goods so pledged).

The special-purpose public warehouse provides the environment required by the type of goods stored: cold storage for perishables, controlled temperature and humidity for volatile liquids, or dehumidified areas for grain or other products that could be damaged by prolonged exposure to humidity.

The *private warehouse* is usually owned outright by the shipper or is occupied on a long-term lease. We can identify two subsidiary types of private warehouses: the *plant warehouse* and the *distribution center.* The plant warehouse stocks inbound raw materials or assemblies used in the manufacturing process, outbound finished goods completed by the factory, and spare parts and cleaning materials used within the company to keep equipment and machinery in good operating condition and the working spaces clean and sanitary. The distribution center emphasizes customer service and fast turnover. Distribution centers, for example, can be located in city areas to group shipments for delivery to a specified neighborhood or even to individual buildings. This use of the center is intended to speed up delivery to customers in congested metropolitan areas and prevent adding to the congestion resulting from numerous parked delivery vehicles.

How Many Warehouses?

When deciding how many warehouses to operate, management faces two constraints: a cost constraint that is based on the cost of building and/or maintaining storage facilities, and a customer constraint compounded of the quality and quantity of service for which the company strives. If both cost and service constraints cannot be met simultaneously, management faces the problem of deciding which constraint should be relaxed. An example will best illustrate this type of managerial decision-making problem.

Suppose a company cannot afford to spend more than $75,000 on any single storage facility. At the same time, an 80 percent level of customer service is desired, using whatever criterion of customer service the company chooses to apply (e.g., errors per 100 orders, number of stockouts per month, number of delayed shipments per year). Since not every warehouse will cost exactly $75,000, we can array the warehouses by number of facilities, cost, and degree of customer service attainable, as illustrated in Figure 6-6.

In this case, management faces the type of trade-off problem described. The $75,000 cost constraint will be exceeded if the 80 percent customer service criterion is to be met since four warehouses will be needed. If the cost constraint is governing in this situation, then only three warehouses can be employed, and the level of customer service achieved will be around 70 percent. In effect, if the customer service criterion can be relaxed, the use of three warehouses will be satisfactory; if the cost constraint can be relaxed, then four warehouses can be used and the 80 percent criterion met.

In this situation we emphasize an important point: Logistics is management, and managerial-type decisions are needed at every step of the way. Nothing is automatic or mechanical, and not nearly enough in a problem situation is predictable. Quantitative techniques and computers can provide an enormous amount of valuable assistance; but when the time comes to pick a course of action, the human manager acts alone. Fortunately, managerial decision making is a skill that can be developed and sharpened with training and practice.

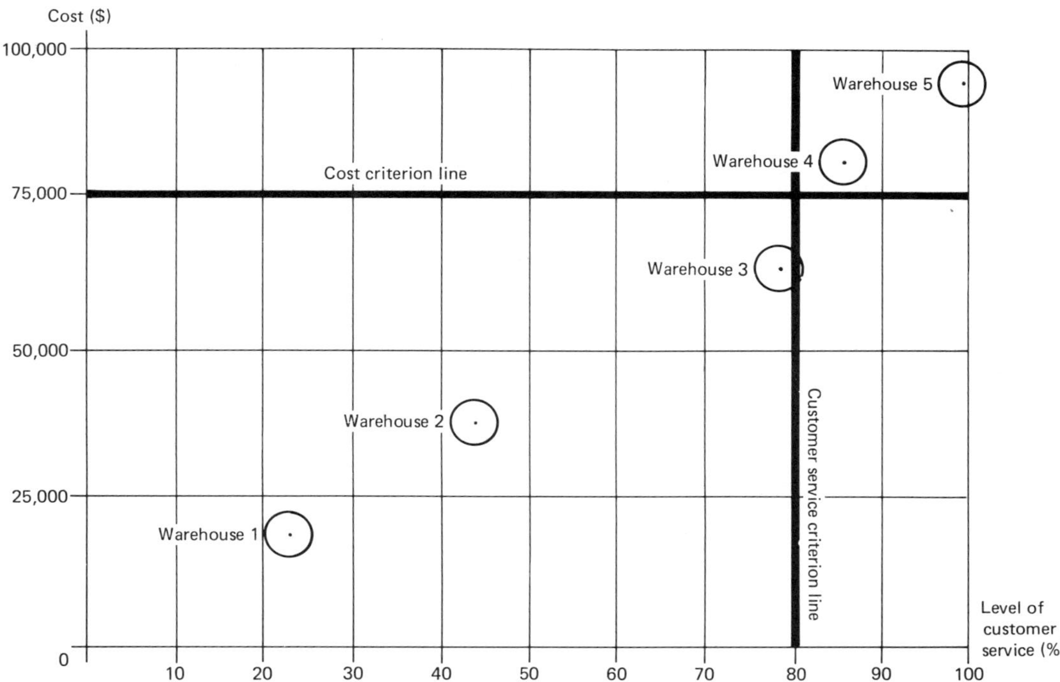

FIGURE 6-6
Optimum number of storage facilities

OBJECTIVES ACHIEVEMENT CHECKUP

The following exercises will help you in assessing the extent to which you have achieved the objectives for this chapter. Where essay-type answers are required, the more you are able to condense your knowledge into a short paragraph of a few sentences the better. Looking back as you work these exercises and rereading parts of the chapter are permissible; but do not look ahead to the answers until you have done your best with the questions.

1. What factors should be considered in siting a manufacturing or storage facility?
2. List the one major advantage and disadvantage of each method of acquiring storage space.
3. You are assigned the task of locating the site at which a distribution center is to be built. Customers and their locations are as follows:

Customer	*Coordinate*	
	East	*North*
A	4	7
B	3	1
C	6	5
D	2	9
E	8	4

Use the center-of-gravity method to determine the coordinates of the point at which you would recommend building the distribution center.

4. Your supervisor prefers that a more precise location for the new distribution center in Question 3 be computed. In addition to the coordinates of customer locations given in Question 3, you also have the following information on quantities shipped to each customer and the transportation costs.

Customer	*Quantity Shipped (units)*	*Transportation Costs*	*Total Costs*
A	5,000	$0.06	$300
B	8,000	0.04	320
C	3,000	0.05	150
D	11,000	0.05	550
E	1,000	0.02	20

5. With the emphasis on precision current in your company, you decide to use the distance–cost method of locating the facility with respect to the location of the nearest competitor. You gather the following information on competitor 2 (consider your company competitor 1):

   Distance between the two competitors: 75 miles
   Production costs: competitor 1, $5 per unit.
   competitor 2, $3 per unit
   Transportation costs: competitor 1, $0.08 per unit
   competitor 2, $0.10 per unit

   And then get to work on the calculations.

6. The general function of storage can be applied to both a warehouse and a distribution center. Differentiate between the two.

ANSWERS TO CHECKUP QUESTIONS

1. Among the siting factors to be considered are:
   a. Type of facility
   b. Availability of material resources, labor, and transportation
   c. Access to markets
   d. Climate
   e. Facilities for employees (e.g., housing, schools, and recreational areas)

2.

Method	*Advantage*	*Disadvantage*
Buy or build	Complete control	Possible underuse
Rent	Flexibility	Short term of the commitment
Lease	Flexibility	Possible underuse during term of lease
Storage in transit	Requires no fixed facility	Requires highly reliable means of transportation

3. Coordinate totals: east, 23; north, 5.2. Dividing these totals by 5 yields the recommended coordinates of the facility: east, 4.6; north, 5.2.
4. First calculate the grand total costs: $1,340. Next compute the location factors for each customer:

	Location Factors	
Customer	*East*	*North*
A	300 × 4 = 1,200	300 × 7 = 2,100
B	320 × 3 = 960	320 × 1 = 320
C	150 × 6 = 900	150 × 5 = 750
D	550 × 20 = 1,100	550 × 9 = 4,950
E	20 × 8 = 160	20 × 4 = 80
	Totals 4,320	8,250

Total transportation costs = $1,340

$$\text{East coordinate} = \frac{4{,}320}{1{,}340} = 3.2$$

$$\text{North coordinate} = \frac{8{,}250}{1{,}340} = 6.2$$

5. If (as it does) $D = C_1 + C_2 = 75$ miles, then

$$C_2 = 75 - C_1$$

The value of C_1 is then calculated as follows:

$$PC_1 + TC_1C_1 = PC_2 + TC_2C_2$$

$$\$5 + 0.08(C_1) = \$3 + 0.10(75 - C_1)$$

$$\$5 + 0.08(C_1) = \$3 + 7.5 - 0.10C_1$$

$$0.08C_1 + 0.10C_1 = \$3 - \$5 + \$7.5$$

$$0.18C_1 = 5.5$$

$$C_1 = 30.6 \text{ miles}$$

From these calculations, we can conclude that your company has the cost advantage from its location to within 30.6 miles of the boundary dividing the market area with your competitor (who has the cost advantage 44.4 miles from the boundary).

6. a. Warehouse: a storage facility whose primary purpose is storage.
   b. Distribution center: a logistical facility whose primary purpose is the movement of goods.

7

Warehousing: Layout and Equipment

CHAPTER OBJECTIVES

Upon completion of this chapter, you should be able to:

A. Comprehend the major design factors in warehouse layout.
B. Understand the value of the unit concept in cost reduction.
C. Appreciate the influence of item turnover rates on warehouse design.
D. Know the packing and marking requirements for handling hazardous freight.
E. Differentiate between the methods of preventing waste of containers and other materials-handling supplies.

What does the word "warehouse" call to mind? A huge building, dark cavernous interior, and goods piled to the roof? Partially correct. But a "warehouse" or storage facility could be a fenced-in yard, paved or otherwise, holding items too large or difficult to fit inside a roofed building. Therefore, when talk turns to warehouses in particular and storage in general, the open, unroofed yard should also be considered.

WAREHOUSE DESIGN

A major factor in the design of a storage facility is the kind of operation intended: mechanical or automated. The mechanized warehouse relies on human labor assisted by materials-handling equipment of various sorts. In the automated warehouse (or distribution center), machines perform work formerly done by human workers. For example, machines in the automated warehouse are able to place items in their storage locations, pick items scheduled for shipment, assemble items in a central location prior to shipment, and keep records.

The turnover rate of the goods stocked will also influence the layout. A low turnover rate (i.e., stock moves rather slowly, remaining in storage for relatively long periods of time) would cause emphasis on the storage function, whereas a high turnover rate (i.e., stock moves in and out of storage rapidly) would cause emphasis on movement and accessibility of the stock.

Design Factors

If accessibility is a prime consideration, each product stored could have a location reserved for it and would always be found in that one spot. Knowing that a certain item can always be found at the same location may be helpful in getting to the stock; but when inventories of that item run low, a great deal of space can be wasted. At the other end of this spectrum, each product could be assigned a storage location as it arrives. Since this method places the item in a different location everytime a replenishment supply arrives, the product can become very difficult to find unless the record keeping is detailed and meticulously accurate. If the most efficient use of space is paramount, this latter procedure would be used.

Use of space, however, involves more than just the area of the floor taken up by the products. Since height is another factor, we are led to the problem of a cube. A cube represents volume (remember, area × height?), and this means both floor space and the area above the floor space must be considered. How can best use of the volume, or cubic area, be achieved?

If best use of cubic volume is to be attained from height (i.e., stacking the items as high as possible), construction costs will probably be lower. A warehouse that obtains its cubic storage capacity from height would take the form of a building that was tall and thin, resting on a small amount of ground area. Offsetting this highly economical use of square footage will be the greater expense of the materials-handling equipment to be used in such a structure. The equipment needed to reach up to pick stock from high stacks is costly. If the cubic capacity is to be obtained from area, construction costs will probably be higher, since more square footage of land area will be needed. This type of building would be large in area but low in height to accommodate the lower stacks of items. The larger construction costs, however, would be offset by the lower cost of the cheaper materials-handling equipment needed to work the low stacks. This type of equipment is both less sophisticated and more durable.

Another factor is the size of the containers to be stored. The warehouse could be arranged by size of containers, with the smaller packages closer to the aisles. Common sense would also dictate placing the faster moving items together in more accessible locations. Packaging that is of an odd size and awkward to handle would also need a more generous allocation of maneuvering room for the materials-handling equipment.

The placement of the loading dock where shipments are transferred to and from transportation carrier equipment determines in good measure the ease with which products can be handled. If a loading dock can be built on opposite sides of a warehouse, goods will be able to pass through in a straight line, as we shall illustrate in a moment. A warehouse, of course, can still function with a single dock on just one side of the structure, but this presents a problem in scheduling, since inbound and outbound shipments could rarely be handled simultaneously without interfering with each other.

In general, a rectangular shape is best when the carriers serving the warehouse are either truck or rail. Since railroad cars present their sides rather than their ends to the platform, a long dock space is needed for side loading and unloading. The rectangle should thus be long and thin. In contrast, trucks back up to the platform for end loading and unloading. In this case, a shorter dock space is needed and the rectangle would approach a square in shape.

The shape and cost of the land area occupied by the facility could determine how the cubic capacity mentioned earlier is to be achieved. Land that is expensive to acquire or small in available area would require a tall, thin building that could accommodate vertical stacking of products to ceiling height. Land that is cheap or large in area would permit building a single-story warehouse with a large floor space and stacks of containers no more than 20 feet high.

In some instances, the nature of the product would give a high priority to order picking. Companies in the mail-order merchandise trade face the problem of picking stock for large numbers of small shipments. Stock pickers, either human or mechanical, should be routed so that stock for as many shipments as possible is gathered in one trip. This would require the grouping of items by frequency of demand, by size or cube, and keeping together families of items that are frequently ordered at the same time. For example, frankfurters and buns, spaghetti and parmesan cheese, envelopes and stationery are all usually ordered at the same time. These pairs of items should be stored in close proximity.

If emphasis is on reducing operating costs, then the ABC system of classification can be used to assign an item to a storage location based on its turnover rate. The highest-volume items could be grouped together as the *A items;* products with a moderate rate of turnover could be grouped together as the *B items;* the *C-items* would be those with the lowest turnover rate.

One common method of reducing handling costs is the use of the *unit concept*. The unit concept requires keeping a shipment as a single unit as long as possible throughout the distribution process. Products can be maintained as a single unit by several methods. One is to bind the shipment on a single pallet and then break down the pallet load only at the last possible step. Shipments made to customers who place large orders can even be kept on the same pallet from the time the pallet is loaded to the time of delivery.

Another method is the use of slip sheets. The items are first en-cased in a bubble of tight-fitting plastic. A heavy sheet composed of paper, plastic, or fiberboard is used as the base of the unit load. Whenever the bubble of goods must be handled, a forklift truck attaches a clamp to the slipsheet and pulls the shipment into position on a truck. The use of the container box also permits keeping the shipment as a unit load locked inside the metal box, which itself is easy to manipulate with materials-handling equipment.

Warehouse Layout

When handling a large number of low-turnover items, storage becomes the main consideration, with accessibility secondary. A warehouse handling this type of product should have wide, deep bays to maximize the

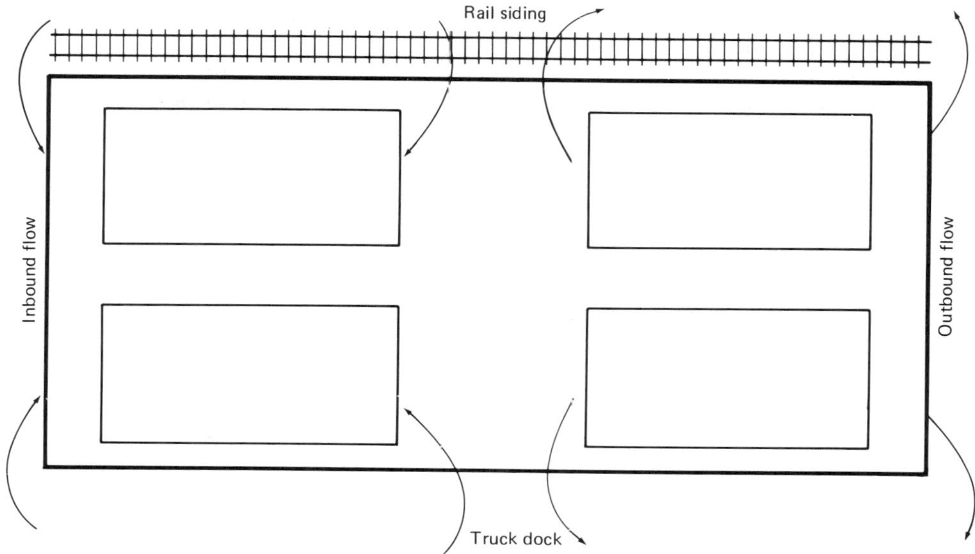

FIGURE 7-1
Storage layout for slow-turnover items

storage area available, high stacks of items, and narrow aisles so that the maximum floor area can be devoted to storage. Such an arrangement might look like the floor plan in Figure 7-1.

In the layout illustrated in Figure 7-1, the truck and rail docks are kept separate by the width of the structure. Railroad tracks, of course, can always be paved over between the rails to permit trucks to use the same area. Scheduling arrivals and departures of equipment would be a difficult problem were that to be done since most often only one mode of transportation could use the dock at a time.

When handling a large number of high-turnover items, speed and ease of access become most important. A warehouse handling this type of product would have shallow bays, low stacks, and wide aisles to permit the materials-handling equipment easy maneuvering and quick access to the products. Such an arrangement might look like the layout illustrated in Figure 7-2. Here, again, the loading and unloading areas for truck and railcars are kept separate. In addition, in both the layouts shown in Figures 7-1 and 7-2, inbound and outbound flows of goods are kept in separate channels to avoid interference between the flows. Separating the working areas for trucks and rail equipment and the inbound and outbound flows also has a safety element considered: Keeping these elements of the situation in separate channels prevents accidents.

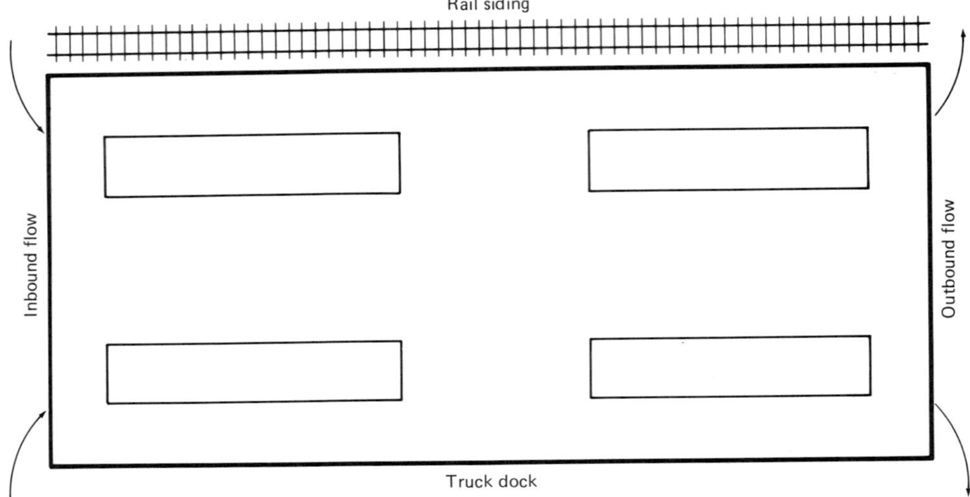

FIGURE 7-2
Storage layout for rapid-turnover items

Should a warehouse be required to handle a mixture of both high- and low-turnover items, some compromises will have to be made in the layout of the warehouse. Storage areas will have to serve both types of items, and this means providing enough room for the materials-handling equipment to work the products and for stacks high enough to use the space to best advantage. An order-assembly area could be provided in a central location in the building to minimize travel to and from the storage locations when orders are being picked and packed. Such an arrangement could look like the layout illustrated in Figure 7-3.

In the event a single loading dock must be used by both rail and truck equipment, the area against the rear wall can be used for additional storage, with the order-assembly area close to the dock rather than being centrally located. Such an arrangement could look like the layout illustrated in Figure 7-4.

Storage and order-assembly areas are perhaps the most important functions for which space must be provided; other functions, however, also demand a share of the facility. For example, space must be provided for the following:

- Parking for employees and visitors.
- A truck waiting area when the dock area is fully occupied.
- Temporary storage for shipments that require repackaging into a size more readily handled by the available equipment.

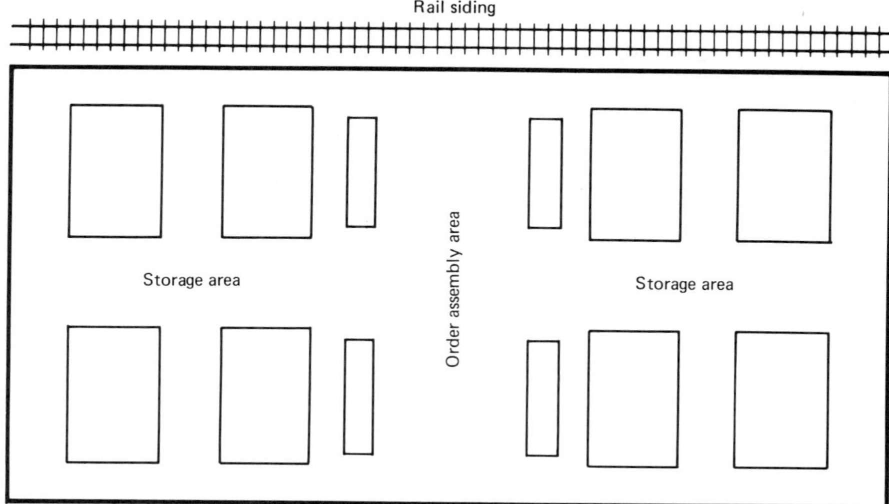

FIGURE 7-3
Storage location for a combination of item types

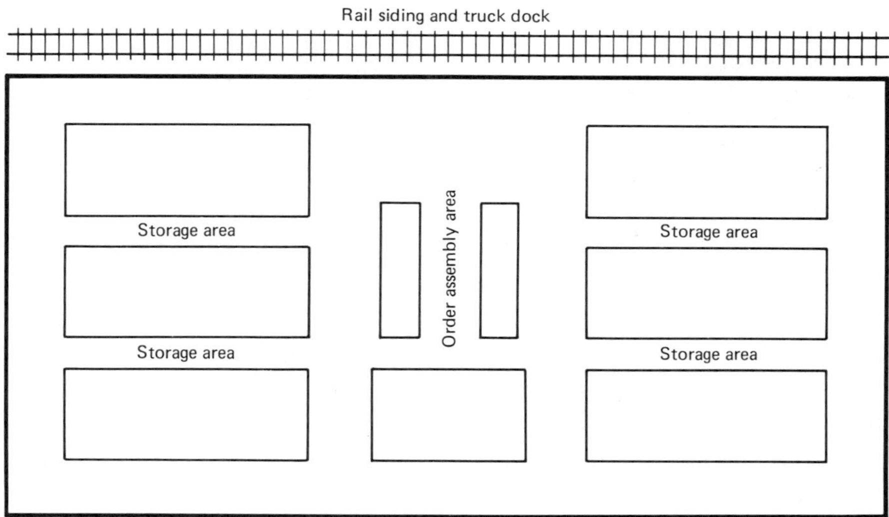

FIGURE 7-4
Storage layout using one side as a rail/truck dock

- Administrative areas, such as:
  - Offices and files
  - Computer equipment (which must often be in a controlled environment to protect the machines from excessive heat, dirt, and vibration)
  - Employee facilities (e.g., lunch, recreation, and rest-room areas, in addition to lockers to store work clothing or uniforms)
  - Special storage areas for:
    - Equipment repair and maintenance
    - Pallet storage
    - Waste collection and disposal
    - Items needing special handling (e.g., perishables that need a chill area, hazardous products that need isolation, or high-value items that need protection and security)
    - Warehouse supplies (e.g., mops, brooms, cleaning materials, and fuel and lubricants for the materials-handling equipment)
- Retail displays if the warehouse also houses a retail outlet or a factory showroom.

Each type of materials-handling equipment used in the warehouse has its own storage and maintenance needs. Such equipment might include computers, computer terminals, and other communications equipment; wheeled pushcarts (like those used in supermarkets); forklift trucks; tractors and trailers; and roller conveyors of various sorts.

Other special factors deserving consideration, but that do not always require the use of floor space, involve employee safety. Safety procedures need to be planned with great care to ensure maximum safety for both employees and freight and to comply with existing federal, state, and local laws. Emergency evacuation routes should be posted in prominent locations throughout the facility. A clear and comprehensive manual of emergency procedures (e.g., what to do in case of fire, earthquake, tidal wave, flood, tornado, or volcanic eruption, if these afflict the locality, and what to do when containers of radioactive materials are found leaking) should be in the hands of every employee. These manuals should be distributed at the very start of an employee's tenure with the company during classroom instruction in the provisions of the manual. To prevent the memory from fading, drills should be held at irregular and unannounced intervals to test the plans.

Hazardous materials always need detailed attention to prevent disasters to personnel and the environment. Hazardous cargo areas need to be separated from other general storage areas and to be so constructed that spills or leaks can be contained in the special area. Bracing and bunkering may be needed to ensure the safety and security of the dangerous substances.

Sanitation is a problem when food products are being handled, not only to protect the freshness of the product, but to prevent infestation by rats and other vermin. Adequate fumigation, though, presents the awkward problem of killing the pests while not contaminating the products stored. Proper sanitation not only protects the product, but also protects the health of the employees who must work in such areas. Goggles, helmets, safety shoes, and protective clothing are all needed, in addition to shower areas to remove any possible contamination from the employee resulting from pesticides or other hazardous materials. Of course, emergency equipment should be available: hoses, fire extinguishers, and alarm systems.

PACKAGING

Since shipments are assembled in distribution facilities, packaging is of importance in the operation of a warehouse. The type of packaging designed by the engineering department must not only meet the provisions of law and the requirements of the carriers who will transport the goods, but it also needs to be designed with the type of materials-handling equipment in mind that will be used to move products around the facility.

Some of the types of equipment used in warehouses to handle stock are conveyors for horizontal movement, cranes for lifting and limited horizontal movement, and various types of trucks, such as forklifts, for lifting and considerable horizontal movement about the floor space or small tractors that tow loaded cars around the warehouse either on rails or rubber tires.

This equipment will travel about the various storage areas of the warehouse in which a variety of packing methods can be used: stacks of pallets; bins, racks, or shelves upon which stock is placed; and gravity flow racks in which packages are stored on inclined conveyors. In this latter type of storage, the removal of one package at the low end of the rack causes a new package to roll into position ready for removal. A knowledge of the method of storage helps in determining the type of packaging needed. A package that will eventually be opened and the contents poured into a storage bin need not be as substantial as a package that will be handled as a unit by different types of materials-handling equipment in different locations for most of its movement through the logistical pipeline.

The type of packaging used for a product is heavily influenced by the characteristics of the product itself. Density is an important factor when handling bulk freight. If the packaging can be reduced in weight, freight charges will be less. We discovered that transportation charges are an important element in determining the boundaries of market areas. A trade-off, however, is present here: Lighter packaging increases the chances for an increase in the frequency of loss and damage.

Related to density is weight and cube. For example, aircraft wing tanks absorb the available cargo space (i.e., the available cube) long before the weight limits of the materials-handling equipment are reached. Handling such freight with a forklift truck presents no problems; but when loading this type of commodity on board a rail box car, the available space is going to be filled completely before the weight-carrying capacity of the car is reached. In contrast, lead ingots will absorb the available weight capacity of a forklift truck or box car long before the cubic space is used. This is also true in the warehouse itself: Heavy, dense products will reach the weight limit of the floor in the warehouse before the cubic space is entirely occupied above the floor area.

Viscosity is a major factor when handling liquids, which flow at different rates depending on the temperature. Some products, like lubricating oil, can be solid at very low temperatures or a light fluid at high temperatures, and become a gas when the temperatures are extremely high. The type of viscosity depends in good measure on the chemical composition of the product and the temperatures to be encountered while in the logistical system. The packaging must be designed to contain the product under all foreseeable circumstances.

Gaseous products, like oxygen or acetylene, are usually stored in compressed form in cylinders or tanks. This requires packaging strong enough to prevent leaking or rupturing in transit and storage.

HAZARDOUS MATERIALS

Hazardous goods require the most careful handling of all. Highly detailed precautions must be taken. These goods, if mishandled, have the capability of harming persons (often fatally) and of damaging property and the environment. Even the product itself requires protection from contamination to retain its purity and be usable by the customer. Hazardous materials are grouped into the following nine classifications:

Class 1: Explosives

Class 2: Gases

Class 3: Flammables and combustible liquids
Class 4: Flammable solids
Class 5: Oxidizing materials
Class 6: Poisons and infectious materials
Class 7: Radioactive substances
Class 8: Corrosives
Class 9: Miscellaneous

One other category of goods must also be identified separately when being shipped by air: magnetized materials that might affect the guidance systems.

Hazardous freight must be identified by special labels on the individual packages and special placards on the transportation equipment that haul these commodities. Some examples of the types of products that must bear these special markings are as follows:

- Explosive materials that can be detonated either by impact, fuse, or chemical deterioration.
- Compressed gases, whose sudden and explosive release may cause harm or damage even though the gas itself is not dangerous.
- Flammables, or gases, liquids, and solids that can be ignited.
- Poisons such as chlorine, which, under proper conditions, can form the deadly war gas phosgene.
- Irritants, such as tear gas or substances that induce vomiting.
- Corrosives such as acids or alkalis.
- Infectious materials, such as bacteria and viruses, that can cause diseases in humans and animals.

Because of the danger of hazardous materials, the regulations concerning their handling are very detailed. These regulations are contained in Title 49 of the Code of Federal Regulations, Title 49 Transportation, Parts 100–199. Included are such provisions as the following:

- The product container, the carrier equipment, and storage areas must be posted with the prescribed labels and placards.
- All handling equipment must be kept scrupulously clean.
- Emergency equipment to be used in case of accident must be readily available and in good working order (e.g., fire extinguishers and hoses, special suppressant chemicals, and such employee protective gear as suits and masks).

- Decontamination equipment must be readily available and in good working condition.
- Warning systems must be installed, working, and tested at prescribed intervals.

The identification of hazardous materials while in storage and transit is made easy by the use of prescribed labels on packages containing them and warning placards on the transportation equipment. An example of a warning label to be placed on a cylinder of a compressed but noninflammable gas is illustrated in Figure 7-5. The diamond label contains the image of the cylinder in which such gases are usually transported, the words NONFLAMMABLE GAS, and the numeral 2 at the bottom, indicating that this material falls into hazardous class 2. A similar diamond-shaped warning placard is affixed to the carrier equipment used to haul this type of freight. Two methods of placarding are in common use: One is the use of the placard plus an identification number displayed on a rectangular orange panel affixed near the placard; the other is the use of a placard with the identification number displayed in the center of the placard, as in Figures 7-6 and 7-7. The identification number indicates the exact substance being transported. In the example in Figure 7-7, the number 1005 indicates a shipment of anhydrous ammonia.

Border of label, gas cylinder, letters, and number in white on a green background.

FIGURE 7-5
Hazardous freight placard

Border of placard, gas cylinder, letters, and number in white on a green background.

1005

Border of panel and numbers in black on an orange background.

FIGURE 7-6
Separate hazardous freight placard and panel

With so many rules, regulations, and provisions of law to keep in mind, some conflicts are bound to occur. For example, protection of the contents of a container must be weighed against the characteristics of the product. A grain or two of radioactive cobalt 60 used in radiation treatments of cancer could be shipped in a small box; but the nature of the material requires a lead-lined barrel with the cobalt 60 container imbedded in cement. Ease of handling thus becomes secondary to protection of the commodity and all who handle it.

TYPES OF PACKAGING

The contents of a package can be indicated on the exterior of the container either in words or symbols. Figure 7-8 illustrates a bar code and a dot code. Exterior markings of this sort are especially important in

Border of placard, gas cylinder, and bottom number in white on a green background. Border and numbers of center panel in black on an orange background.

FIGURE 7-7
Combined hazardous freight placard and panel

identifying shipments if the packages happen to get separated from the accompanying shipping papers.

Perhaps the most awkward trade-off is between heavy and light packaging in those instances in which the type of product permits a choice. Using packaging heavier than might be required by the transportation carrier or by law has the advantage of providing greater protection from damage, but at greater cost. In fact, if the number of damage claims starts approaching zero, the chances are excellent that the product is being overpackaged. A lighter packaging results in lower costs, but risks greater exposure to damage in transit. Thus, a very high number of damage claims probably indicates that the product is being underpackaged.

The ABC method of classification might assist in controlling the costs of packaging. Type A items could be those with many unknowns in the logistical situation. The exact nature of conditions at destinations may not be known, nothing can be learned about the type of storage available en route and at destination, or conditions aboard the carrier may not be known with certainty. Such items would require the maximum protection and the heaviest packaging. Type B items could be those with fewer unknowns in the logistical situation. The packaging

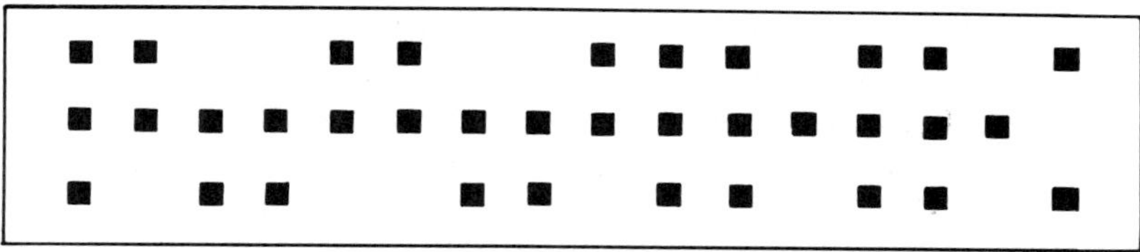

Dot code

Bar code

FIGURE 7-8
Examples of dot and bar codes

should be sturdy enough to withstand the probable conditions en route, but need not be as heavy as the packaging for type A items. Type C items could be those about which considerable information is available concerning conditions in transit and could withstand the lightest acceptable packaging consistent with protecting the product.

What happens when the engineering department comes up with a new type of packaging, or a new product appears for which existing packaging requirements are not quite suitable or not even specified in carrier tariffs? Freight classification tariffs are often very specific about the construction of the containers used in shipping certain products. Logistical managers, however, can apply for special six-month exemptions from the tariff specifications to test new types of packaging. For example, severe problems were caused by the substitution of plastic bottles for glass containers. Glass, of course, can shatter on impact but is still strong enough to withstand stacking. Plastic bottles do not shatter, but will buckle if stacked too high. Considerable experimentation followed to determine just how the new plastic containers should be stored and transported.

Functions of Packaging

Packaging has two basic functions to perform. One is to assist the marketing department in selling the product. If the item is shipped in customer-sized packs (e.g., individual tubes of toothpaste), the type of packing will influence the type of transportation package into which the smaller packs are loaded for storage and shipment. If the customer-sized packs are very sturdy, the exterior transportation packing can be quite light; but if the customer-sized containers are fragile (e.g., the cardboard containers in which eggs and light bulbs are shipped), the transportation packaging must be substantial.

The marketing department may wish to have display advertising printed on the exterior packaging since so many of the new discount food supermarkets use the transport packing as bins from which customers pick their choices of products. Unfortunately, the display material may also obscure the transportation markings. Some close cooperation will thus be required with the marketing personnel to prevent a fancy exterior design from interfering with the movement of the shipment. The package, if possible, should also deter shoplifters. Have you ever wondered why small watch batteries are packed in a plastic bubble attached to a large card? The bubble is so firmly affixed to the card that any effort at opening the bubble would be obvious to observers. The large card makes the item more difficult to slip into a pocket and is a further deterrence to all but the most determined thief.

The other major function of packaging is protective in nature. To meet this protection requirement, the packaging should:

- Keep the items enclosed and separate from other items.
- Prevent the contents from shifting around in transit.
- Cushion the contents from vibration and shock.
- Support the weight of other containers in a stack.
- Distribute the weight of the contents evenly throughout the container.
- Provide enough exterior surface on which to attach labels or placards.

Packaging Requirements in Transportation

Carriers establish packaging requirements both to make best use of their equipment and to reduce loss and damage claims. In consequence, packaging is frequently specified in the freight classification tariffs, which then give these requirements the force of statutory law. The freight tariff might specify that a particular commodity be shipped in bales compressed to the stated number of pounds per square foot or in containers of specified strength. In addition to being very precise about the type and strength of shipping containers, the tariffs also specify how products may be shipped when not in containers. Tariffs often specify that commodities may be shipped flat, set up (usually abbreviated as SU in the tariff), or knocked down (abbreviated as KD).

Containers must be clearly labeled with the contents of the packaging. This can be done with words or abbreviations, in coded words, or by the use of the dot or bar codes illustrated in Figure 7-8. These codes can be read by optical scanners the same as those used in supermarket checkout counters. When containers are loaded aboard the equipment of a carrier, the tariffs may also require that the shipment be braced against movement. This bracing can be constructed of wooden boards (called dunnage) or even inflatable bags to prevent the freight from shifting around the equipment and becoming damaged or damaging other freight.

The box container has become especially useful as a means of shipping quantities of smaller packages or high-value cargo. The box container came into wide usage by the military logistical system in which it was called a CONEX box, taken from the method of shipping military freight known as Container Express. The box is simply a metal container 8 feet wide, 8 feet high, and from 10 to 20 feet long. Since the items to be shipped are loaded in the box, relatively light packaging can be used for the contents. The ease with which the metal box containers can be handled in both storage and transit has made them popular. In fact, entire ocean-going vessels have been fitted to handle only containerized cargo.

The configuration of the shipment might also compel attention to other special considerations. Bridges and roads may impose limitations on height, weight, and width. Overhead structures such as viaducts and power or telephone lines can be highly restrictive of the size of shipping containers. Many political jurisdictions pose strenuous objections to endangering the lives and property of their citizens with shipments of oversized or hazardous materials.

Preventing Waste

The use of wood in constructing pallets and in the dunnage used to block and brace freight can be extremely wasteful if not carefully supervised. The boards used in blocking and bracing should be removed from the carrier equipment after the shipment has been unloaded, including the boards that appear to have been damaged in transit by breaking or splintering. Such damaged boards can be used to repair or build new pallets, while the boards that are still intact and serviceable should be preserved for use in other shipments.

The same is true of fiberboard containers. Fiberboard cartons can be reused many times unless badly torn. Common sense needs to be exercised here. A carton used to ship materials that could damage or contaminate other freight for which the box might be used should be discarded. But if such a carton is discarded, should it be destroyed or recycled? Boxes contaminated with radioactive materials obviously cannot be recycled (no, contact with them will not cause you to glow blue in the dark), but disposal presents other problems. Burning such containers spreads radioactive contamination in the smoke and burial creates another set of problems. Federal, state, and local governments all have regulations controlling the disposal of contaminated wastes. The occasional incident of the ancient waste site in which harmful substances were buried years previously comes to mind when it now oozes back into the light of day. Expect control regulations to be more frequent and stringent in the future.

The "kepone disaster" of 1975 involving the Allied Chemical Company illustrates this point. Two former Allied Chemical employees started up an operation manufacturing the pesticide kepone in a converted gas station under contract to the main plant at Hopewell, Virginia. This contract operation contaminated the nearby James River to the extent that the river was closed to fishing. The seafood industry in the Chesapeake Bay area suffered hundreds of millions of dollars in damage, not to mention the people who were sickened by the chemical. A federal grand jury in Richmond returned an indictment containing 1,104 counts against the company, the contractors, and the city of Hopewell. The end result was an informal partnership between the company and the government (in the form of the Environmental Protection Agency) to establish standards and procedures to prevent such an incident from ever happening again. By 1980, the EPA was praising the company for its spirit of willing cooperation.

This incident is an example of the cooperative nature of waste disposal activities. Even when waste material is not contaminated with hazardous substances, the problem of disposal is still present. One of the better courses of action is to prevent the waste from forming through reuse or recycling of the packaging materials. If the waste materials cannot be reused in some fashion, then disposal should be made in strict compliance with existing local and federal laws even though that may mean incurring a considerable cost.

These waste disposal activities should be carried on in full and open cooperation with the local enforcement agencies. Plans, procedures, regulations, sites—all need a great deal of coordination with the responsible public authorities. In fact, this sort of open, though often informal, cooperation with the government can assist in the amicable solution of other problems that originate in commercial and industrial operations: noise, dirt, liquid and gaseous emissions, unsightly piles of materials, and a host of other problems that have potential for creating friction and poor relations with the surrounding community.

WAREHOUSE PRODUCTIVITY

Accounting and financial analysts frequently use data in the form of various ratios to form judgments in their technical areas. Similar mathematical relationships can be used to evaluate productivity in warehouse operations. For example, suppose a manager wanted to judge how well space was being used in the warehouse. Storage and aisle ratios would be helpful. The storage ratio would provide information on how much of the total space in a storage facility was actually being used for storage. This ratio can be computed using the following equation:

$$\text{Storage ratio} = \frac{\text{storage area}}{\text{total area}}$$

An aisle ratio would provide similar information concerning the proportion of the total floor space devoted to such nonstorage activities as aisles:

$$\text{Aisle ratio} = \frac{\text{aisle space}}{\text{total area}}$$

The efficiency with which the work force employed at the warehouse performs can be judged by a similar process. For example, the proportion of time devoted to receiving goods can be computed as follows:

$$\text{Receiving ratio} = \frac{\text{quantity received (in weight or amount)}}{\text{total available labor hours}}$$

The proportion of work time devoted to order picking can be computed using the following equation:

$$\text{Order picking ratio} = \frac{\text{orders picked}}{\text{available labor hours}}$$

One persistent problem in warehouse management is the assignment of shipping priorities. This involves determining how urgent an order may be and the order to be observed in shipping out products. An answer to the question "When should an order be shipped?" is provided by the critical ratio. This ratio measures the urgency of an order, which certainly is a matter to be considered in deciding the order in which shipments should be made. The critical ratio is computed by using the following equations:

$$\text{Critical ratio} = \frac{\text{days of supply}}{\text{lead time remaining}}$$

where

$$\text{Days of supply} = \frac{\text{stock on hand} - \text{safety stock}}{\text{average daily demand}}$$

Recall that *lead time* is the time elapsing between the receipt of an order and shipment to the customer. The criteria to be used in judging the urgency of an order are as follows:

- If the critical ratio is greater than 1, the order is ahead of schedule and can be delayed if so desired until the regular shipment date.
- If the critical ratio equals 1, the order is on schedule.
- If the critical ratio is less than 1, the order is behind schedule and should be shipped next.

An example will illustrate the use of the critical ratio. Suppose the following information is known concerning a specific shipment:

Lead time is normally 18 days.

Stock on hand: 125 items.

Safety stock: 75 items.

Average daily demand: 25 items.

If today is the twelfth day in the lead-time sequence for this order, the critical ratio will be

$$\text{Days of supply} = \frac{125 - 75}{25} = \frac{50}{25} = 2$$

$$\text{Critical ratio} = \frac{2}{6} = \frac{1}{3}$$

Since the critical ratio is less than 1, we can conclude that the order is behind schedule and should be shipped out by the next available transportation.

Once the efficient operation of the warehouse or distribution center has been assured, the next problem is to gain control over the inventory. Inventory control and management will be addressed in the following two chapters.

OBJECTIVES ACHIEVEMENT CHECKUP

The following exercises will help you in assessing the extent to which you have achieved the objectives for this chapter. Where essay-type answers are required, the more you are able to condense your knowledge into a short paragraph of a few sentences the better. Looking back as you work these exercises and rereading parts of the chapter are permissible; but do not look ahead to the answers until you have done your best with the questions.

1. What are the major factors that should be considered in the design of a warehouse layout?
2. Explain how the unit concept contributes to cost reduction in the operation of a storage facility.
3. How would you alter the design of a warehouse for low- or high-turnover items?
4. How do the marking requirements for hazardous freight differ from those of nonhazardous products?
5. List the different methods by which the waste of containers and handling supplies can be prevented.
6. Your warehouse supervisor reports that shipments are beginning to get behind schedule. He has three shipments in process, but the carrier will be able to take only one of these. The following information is available on three shipments:

Data	*Shipment 1*	*Shipment 2*	*Shipment 3*
Lead time	10 days	15 days	12 days
Stock on hand	200 units	75 units	175 units
Safety stock	25 units	10 units	20 units
Average daily demand	30 units	15 units	5 units
Day in lead-time sequence	6th	10th	8th

Using the critical ratio, in what order would you schedule the three shipments for dispatch?

ANSWERS TO CHECKUP QUESTIONS

1. a. Type of operation, whether mechanized or automated.
   b. Turnover rate of the inventory stock.
   c. Accessibility of stock.
   d. Available volume or cubic area.

e. Placement of loading dock.
f. Size and shape of containers in which stock is packed.
g. Cost of land.

2. Keeping a shipment as a single unit as long as possible in the logistical process reduces the cost of handling by eliminating small batches of units, which are more costly to handle. Transportation costs are also lowered since shipments can be made in larger unit load quantities.
3. a. High-turnover items require ease of access and speed in picking and packing stock. This type of product requires a warehouse with shallow bays, low stacks, and wide aisles.
   b. Low-turnover items require emphasis on storage. This type of product require a warehouse with wide, deep bays, high stacks, and narrow aisles to permit maximum use of storage space.
4. Placards and labels must be used on shipments of hazardous products to indicate both the type of product and the nature of the hazard (e.g., explosives, poisons, corrosives).
5. a. Promptly remove all blocking and bracing materials from carrier equipment.
   b. Leave no blocking and bracing materials lying about the loading/unloading area.
   c. Use damaged materials wherever possible to repair boxes, crates, and pallets.
   d. Reuse fiberboard containers that are still serviceable (e.g., in good condition, not contaminated).
   e. Reuse rather than destroy materials.
6.

Item	*Shipment 1*	*Shipment 2*	*Shipment 3*
$\text{Days of supply} = \dfrac{\text{stock} - \text{safety stock}}{\text{average daily demand}}$	$\dfrac{175}{30} = 5.8$	$\dfrac{65}{15} = 4.3$	$\dfrac{155}{5} = 31$
$\text{Critical ratio} = \dfrac{\text{days of supply}}{\text{lead time remaining}}$	$\dfrac{5.8}{4} = 1.5$	$\dfrac{4.3}{5} = 0.9$	$\dfrac{31}{4} = 7.8$

Based on these critical ratios, the shipments should be readied for dispatch in the following order:

a. First priority: shipment 2, whose critical ratio of 0.9 indicates that the order is behind schedule.
b. Second priority: shipment 1, whose critical ratio of 1.5 indicates that the order will soon be behind schedule unless dispatched.
c. Third priority: shipment 3, whose critical ratio of 7.8 indicates the order is well ahead of schedule.

8

Inventory Management: Forecasting

CHAPTER OBJECTIVES

Upon completion of this chapter, you should be able to:

A. Calculate forecasts by semiaveraging and exponential smoothing.
B. Know the capabilities and limitations of forecasting.
C. Understand the necessity of assessing the accuracy of forecasting techniques.
D. Use the inequality coefficient to track the accuracy of forecasts.
E. Appreciate the need for using a variety of forecasting techniques.

"What is that pile of goods in the warehouse?"

"Why, that's our inventory."

"What's 'inventory'?"

The best way to start a new topic is to define the basic terminology. Therefore:

> INVENTORY: A stock of goods intended for use in a manufacturing process, resale, internal consumption within an organization, or for maintenance.

A number of problems present themselves in the management of inventory, none of which are especially simple. We shall consider these in this and the following chapter. One complicating factor is that inventory represents both a cost and an asset to the company.

As much as business enterprises wish to reduce their costs, the cost aspect of inventory is not subject to minimization simply by holding the smallest possible inventory. Too small an inventory risks frequent stockouts, and stockouts leave the impression that the firm is an unreliable supplier. Stockouts actually add to costs through the administrative action required to backorder goods for those customers who are willing to wait and to cancel orders already placed by customers who are not willing to wait. An interesting speculation concerns the procedures for determining accurately the number of impatient customers who never place another order, but turn permanently to another source of supply.

From the viewpoint of an asset, the opposite side of the coin applies if inventories are too large. Stockouts can be virtually eliminated by carrying a large enough inventory; but too large an inventory poses additional problems. Inventories of any size must be stored somewhere, insured, accounted for, and frequently taxed. The larger the inventory, the more severe these problems become. Large inventories tie up funds that could be used to better advantage elsewhere in the organization. Worst of all, these funds are subject to shrinkage caused by obsolescence, depreciation, pilferage, interest charges, and taxes. All these considerations erode the positive asset nature of inventory. To these problems must be added the normal operating costs connected with inventory management.

For convenience in summarizing, let's divide inventory costs into four categories and list some of the items included in each category as follows:

1. Order costs
   - A. Cost of placing an order with a vendor of materials:
     - (1) Preparing a purchase order

      (2) Processing payment
      (3) Receiving and inspecting the materials
   B. Ordering from the plant:
      (1) Machine setup
      (2) Start-up scrap generated from getting a production run started
2. Inventory carrying costs
   A. Costs connected directly with the materials:
      (1) Obsolescence
      (2) Deterioration
      (3) Pilferage
   B. Financial costs:
      (1) Taxes
      (2) Insurance
      (3) Storage
      (4) Interest (as the cost of capital borrowed to acquire and maintain the materials)
3. Out-of-stock costs
   A. Back ordering
   B. Lost sales
4. Capacity costs
   A. Overtime payments when capacity is too small
   B. Layoffs and idle time when capacity is too large

How much inventory to carry is a question with important financial—and profit—overtones. The answer to this problem is one of the most difficult faced by any company. The first step toward a solution is forecasting the future. Unfortunately, even the wisest and most knowledgeable forecaster peers into a murky crystal ball. If forecasting is somewhat less than accurate at times, why even bother? Because management cannot afford to be overtaken continually by the course of events. Forecasting assists in charting a course for the future, and for that, preparation must be made by selecting the forecasting techniques that work the best for the company. For the moment, let's try our hand at peering into that murky future.

FORECASTING

The shelves are full of books on forecasting techniques, and computer routines trip by in endless profusion, some fairly simple, others intensely mathematical. To get some idea of how forecasting is done, we will examine two common, useful techniques: semiaveraging (also called trend analysis) and exponential smoothing. The accuracy of a forecasting technique is a difficult problem to overcome since the only information available to the forecaster is past data. A forecasting method thus is only as accurate as the expectation that the past is a good index to the future.

If operations are going along smoothly, the most naive forecast imaginable is probably as accurate as any: Tomorrow (next week, next month, or whatever time period is being used) will be like today. If, in contrast, the business climate is stormy and the commercial tides produce irregular waves, then the accuracy of the forecast deteriorates. Perhaps the best advice to a forecaster or manager is to keep checking up on the forecast. The moment the forecast diverges too widely from a predetermined standard of accuracy, corrective action should be taken, perhaps by trying another method.

SEMIAVERAGING

Semiaveraging (or trend analysis) is occasionally used as a substitute for the more laborious linear regression analysis. Essentially, least squares linear regression attempts to fit a straight line through the center of gravity of a series of plot points on a graph. Linear regression computer routines are ideal for performing the computations and, if the ups and downs in the data are not too pronounced, can produce useful results. Semiaveraging provides an approximation of the regression results with far fewer calculations. The calculation equation is as follows:

$$F_t = \overline{Y}_1 + \frac{\overline{Y}_2 - \overline{Y}_1}{x} X$$

where F_t = forecast

$\overline{Y}_1$ = arithmetic average (the mean) of the first data group

$\overline{Y}_2$ = arithmetic average of the second data group

x = code number of the middle data point of the second data group

X = code number of the period to be forecast

The procedure (and explanation of some of the preceding definitions) is as follows:

- Divide the set of data points (data for each year, week, month, or time unit being used) into two equal groups.

- If the set of data points includes an odd number of points, ignore the middle data point to form equal groups around the center.
- Calculate the mean for each of the two groups of data points.
- Each data point in the set must now be assigned a code number, even the data point being forecast.
- In the coding process, assign zero (0) to the data point at the center of the first group.
- Use minus numbers to code the data points above the 0 point; use plus numbers to code the data points below the 0 point.

But suppose the first group of data points consists of four entries and does not have a middle entry? Then do the following:

- Assign code numbers +1 and −1 to the data points in the center of the group as follows:

Code	*Data*
−2	104
−1	96
+1	93
+2	101

And continue on with the procedure:

- *Don't forget this:* Even though the center data point in the set is omitted when forming the two equal groups, that center data point must still be included in the order of coding.
- *And don't forget this either:* Continue the order of coding beyond the last data point (including the points for which no data exist) up to and including the time period being forecast.
- Now substitute the required numbers in the calculation equation to produce the desired forecast.

If all this sounds miserably confusing the first time around, the procedure is simple enough to become boring very rapidly when repeated a few times. Since the best way to learn a calculation routine is to try it, let's try a forecast using semiaveraging.

Suppose a company has gathered data concerning actual sales from January through July. Management wants a forecast of sales for September to be used as a basis for determining the level of inventory that should be maintained. The actual sales figures are as follows:

Code	*Month*	*Sales*	
−1	January	115	First group
0	February	110	
+1	March	90	
+2	April	100	
+3	May	115	Second group
+4	June	120	
+5	July	125	
+6	August	Not required	
+7	September	?	

Now follow the procedural steps.

- Divide the set of data points into two groups, omitting the April data to form the two equal groups.
- Code each month, assigning 0 to the February data.
- Code months in strict sequence, even the April data omitted to form the two equal groups.
- Include in the coding the month omitted between July and the forecast month of September.

Next compute the simple arithmetic averages of the sales figures in the two groups:

$$\overline{Y}_1 = \frac{115 + 110 + 90}{3} = 105$$

$$\overline{Y}_2 = \frac{115 + 120 + 125}{3} = 120$$

All that remains now is to substitute numbers into the calculation equation to obtain the forecast for September:

- Substitute the group averages in the equation.
- Use the code number of the middle month (June) of the second group for x.
- Use the code number for the forecast month (September) for X.

$$F_t = \overline{Y}_1 + \frac{\overline{Y}_2 - \overline{Y}_1}{x} X$$

$$= 105 + \frac{120 - 105}{4} 7$$

$$= 132$$

But what if the second group of data points has an even number without a middle point? Simply average the code numbers on each side of what would have been the middle code number as follows:

Code	*Month*	
+5	October	
+6	November	Average the code numbers on
+7	December	each side of the middle
+8	January	

Thus the average to be used as the code number of the "middle" point is +6.5, which will be used as x in the equation.

For those of a graphic rather than a mathematical turn of mind, a semiaveraged forecast can also be produced visually. Use the Y axis of the graph to measure the quantity of sales and the X axis to measure time, as in Figure 8-1. Then plot the average sales for each of the two groups over the central time period of each group. Connect the two

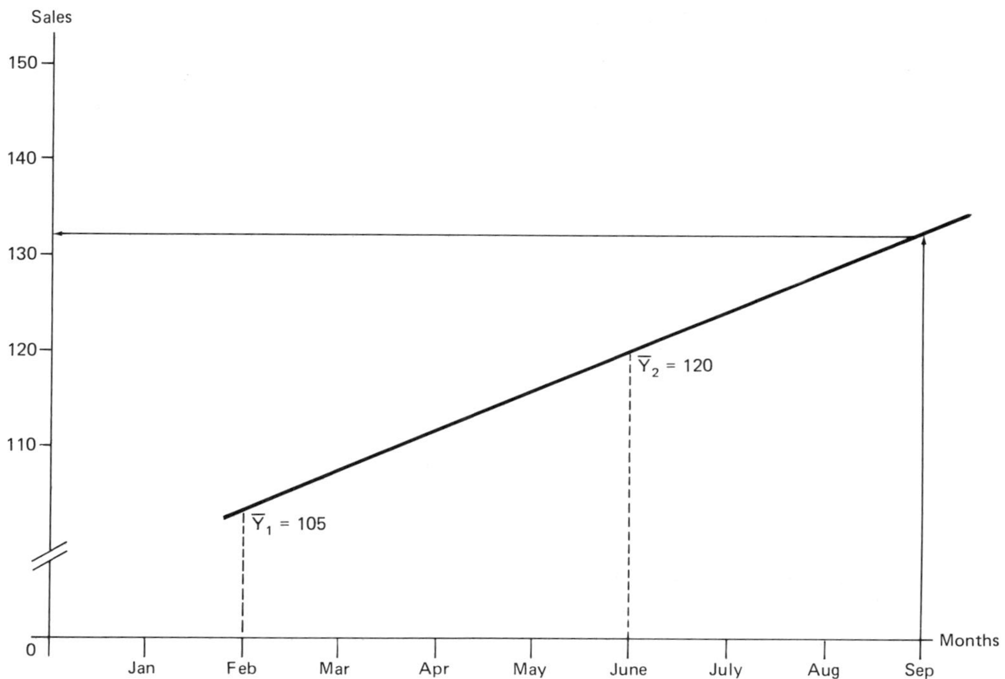

FIGURE 8-1
Plotted semiaveraged trend line

average points plotted on the graph with a straight line (the trend line). The forecast for September is then produced by going up from the September point on the X axis to the trend line, and then directly to the left to the Y axis. The forecast for September can then be read directly, where we end up on the Y axis, and should be the same as the figure computed for September using the calculation equation.

EXPONENTIAL SMOOTHING

The other method of forecasting considered here actually combines two other methods. As a result, exponential smoothing is a technique that is really a weighted moving average.

> MOVING AVERAGE: An arithmetic average calculated by removing the oldest data point each time a new data point is added to the calculation.
>
> WEIGHTED AVERAGE: An average that multiplies each data point by a predetermined numerical weight before proceeding with the arithmetic involved in averaging (the sum of the numbers used as weights is used as the divisor). Greater or lesser emphasis can be given the old or the more recent data, depending on which end of the array of data points to which the weights are applied.

Exponential smoothing combines both types of averages in the following calculation equation:

$$F_t = F_{t-1} + \alpha (A_t - F_{t-1})$$

where F_t = new forecast

F_{t-1} = previous forecast

α = weighting factor assigned a value between 0 and 1 (usually between 0.1 and 0.3)

A_t = actual current period data

Thus exponential smoothing arrives at a forecast as a proportion of the forecast for the previous period plus a fraction of the forecast error (i.e., the difference between the forecast and actual results).

Exponential smoothing works reasonably well when forecasting individual items of inventory and detects trends readily. The choice of α, however, is critical. For example, a small α gives greater weight to the old forecast (by subtracting a smaller figure for the error); the higher the α, the greater the weight given to the actual data. Giving greater weight to the actual data, however, risks having the procedure react too quickly and give erratic forecasts; giving too much weight to the old forecasts risks too slow a reaction and forecasts that appear deceptively smooth. If a firm has been using moving averages as a forecasting technique, the α to be used can be calculated with the following equation:

$$\alpha = \frac{2}{n+1}$$

where n is the number of time periods used in the moving average.

To avoid using F_{t-1} twice in the process of calculation, we can derive a more efficient formula as follows: Take the original equation,

$$F_t = F_{t-1} + \alpha (A_t - F_{t-1})$$

and remove the parentheses:

$$F_t = F_{t-1} + \alpha A_t - \alpha F_{t-1}$$

Now rearrange the terms:

$$F_t = \alpha A_t + F_{t-1} - \alpha F_{t-1}$$

Then rewrite the equation to simplify the task of calculation:

$$F_t = \alpha A_t + (1 - \alpha) F_{t-1}$$

The results to be obtained from this technique can best be illustrated with another example. Suppose the calendar tells us that it is now the end of January and we want to forecast sales for February. Past experience indicates that 0.3 works best as the α smoothing constant. Someone during the preceding December has forecast a sales figure of 75, which we will use initially as the F_{t-1}. The forecast turned out to be exactly correct with sales at 75 in January, which we will use initially as the A_t.

At the end of January, our forecast for February will be

$$F_t = \alpha A_t + (1 - \alpha)F_{t-1}$$

where F_t = forecast for February

A_t = actual sales in January

F_{t-1} = forecast for January (made in December)

Or, by substituting data into the equation,

$$F_t = 0.3(75) + 0.7(75)$$
$$= 75$$

At the end of February, we want a sales forecast for March. Although 75 units were forecast for sales in February, actual sales for that month turned out to be only 70 units. The forecast for March will be

$$F_t = 0.3(70) + 0.7(75)$$
$$= 73.5$$

At the end of March, we will want a sales forecast for April. Although 73.5 units were forecast as sales for March, actual sales for that month turned out to be 75 units. The forecast for April will be

$$F_t = 0.3(75) + 0.7(73.5)$$
$$= 74$$

A sales forecast for May, if actual sales during April turned out to be 80 units, would be

$$F_t = 0.3(80) + 0.7(74) = 75.8$$

And so on until exponential smoothing begins to give unacceptably bad forecasts.

With shelves full of forecasting techniques, which should we use? Try a few that look promising and can be computed either manually or with the computer equipment available in the company. Select data for ten previous time periods. Use the data in the first nine periods with the selected forecasting methods to predict the tenth period; then use the technique that gave the most accurate forecast for the tenth period. How does a manager or forecaster know when the method in use is not forecasting well? The answer is to track the forecast continuously.

TRACKING THE FORECAST

Users and makers of forecasts are always advised to keep careful track of the accuracy of the method being used. Charts can be used to determine when the forecasts deteriorate beyond an acceptable point. The method we will use to illustrate assessing the accuracy of a forecasting technique over time is the inequality coefficient. The inequality coefficient as used here compares the mean-squared error (MSE) of the forecasting technique being used and the mean-squared error of a forecast of "No change." Any two forecasts can be compared by this method, but for our purposes here we will use the forecasts just generated by exponential smoothing and the most naive of all forecasts, "No change." Let's tabulate actual and forecast sales data and square the errors experienced by both forecasting methods. Forecast errors are computed as follows:

Forecast error = forecast data − actual data

The tabulation of the preceding data would appear as follows:

	Sales		*Exponential Smoothing*		*No change*	
Month	*Actual*	*Forecast*	*Error*	*Error²*	*Error*	*Error²*
January	75	75	0	0	—	—
February	70	75	5	25.0	−5	25
March	75	73.5	−1.5	2.25	5	25
April	80	74	−6.0	36.0	5	25
May	85	75.8	−9.2	84.64	5	25
			Totals	147.89		100

The mean-squared errors would be calculated as follows:

$$\text{(MSE of the exponentially smoothed forecast)} = \frac{1}{n}\sum(\text{forecast} - \text{actual})^2$$

$$= \frac{147.89}{5}$$

$$= 29.578$$

$$\text{(MSE of the ``No change'' forecast)} = \frac{1}{n}\sum(\text{forecast} - \text{actual})^2$$

$$= \frac{100}{4}$$

$$= 25$$

The equation for calculating the inequality coefficient is as follows:

$$U^2 = \frac{(1/n)\sum(\text{forecast} - \text{actual})^2 \text{ (for the method in use)}}{(1/n)\sum(\text{forecast} - \text{actual})^2 \text{ (for the criterion method)}}$$

$$= \frac{\text{MSE (for exponential smoothing)}}{\text{MSE (for ``No change'' forecast)}}$$

$$= \frac{147.89}{25}$$

$$= 5.9156$$

The inequality coefficient is interpreted as follows:

- If U^2 is less than 1, the current forecasting method is producing better results than the reference criterion forecast.
- If U^2 is greater than 1, the current forecasting method is producing results inferior to the reference criterion forecast.
- If U^2 is exactly equal to 1, both methods are producing equally satisfactory results.

In the last case we must face the conclusion that we have been wasting our time with exponential smoothing and had better look for a new forecasting technique. As an alternative, we might want to experiment (if we have the time) and see if a different smoothing constant might give better results.

Having now given some attention to predicting what inventory might have to be, we can now turn our attention to the various methods of controlling the inventory. These techniques should help the manager to hold down the sheer size of the inventory (with the attendant cost savings) and to provide reasonable insurance against stockouts (with even more attendant savings). Turn to the next chapter for this.

OBJECTIVES ACHIEVEMENT CHECKUP

The following exercises will help you in assessing the extent to which you have achieved the objectives for this chapter. Where essay-type answers are required, the more you are able to condense your knowledge into a short paragraph of a

few sentences the better. Looking back as you work these exercises and rereading parts of the chapter are permissible; but do not look ahead to the answers until you have done your best with the questions.

1. Given the following monthly demand data for the first six months of the year, calculate a forecast for November using semiaveraging:

Month	*Demand*
January	30
February	25
March	15
April	20
May	30
June	35

2. Using the demand data given in Question 1, forecast demand for July using exponential smoothing. The following additional information will be helpful:
   a. Use a smoothing constant of 0.2.
   b. Actual sales in June were 35, as shown in the table, but had been forecast for 30.
3. What do you think is the major weakness of most forecasting techniques?
4. You suspect the forecasting method now in use to predict demand is not as accurate as you would like. Use the inequality coefficient to determine whether or not the current forecasting technique is better than a simple forecast of "No change." The following information is available:

	Demand	
Month	*Actual*	*Forecast*
March	120	115
April	130	135
May	135	130
June	135	130
July	130	125

5. How would you select the best forecasting method for your company?

ANSWERS TO CHECKUP QUESTIONS

1.

Code	Month	Demand	
−1	January	30	
0	February	25	$\overline{Y}_1 = \frac{70}{3} = 23.3$
+1	March	15	
2	April	20	
3	May	30	$\overline{Y}_2 = \frac{85}{3} = 28.3$
4	June	35	
5	July	—	$F_{\text{Nov}} = 23.3 + \frac{28.3 - 23.3}{3} 9$
6	August	—	
7	September	—	$= 23.3 + \frac{5}{3} 9 = 23.3 + 15$
8	October	—	
9	November	?	$= 38.3$

2.

$$F_{\text{Jul}} = 0.2(35) + 0.8(30)$$
$$= 7.0 + 24$$
$$= 31 \text{ units}$$

3. Probably the leading contender for the major weakness of most forecasting methods is the necessity of relying on past data as a basis for the forecast. The forecast will be in error to the extent that the past is a poor indicator of the future.

4.

	Demand		Current Method		"No Change"	
Month	Actual	Forecast	Error	Error^2	Error	Error^2
March	120	115	−5	25	—	—
April	130	135	+5	25	+10	100
May	135	130	−5	25	+5	25
June	135	130	−5	25	0	0
July	130	125	+5	25	+5	25
			Totals	125		150

$$\text{MSE} = \frac{1}{n}\sum(\text{forecast} - \text{actual})^2$$

$$\text{MSE (current method)} = \frac{125}{5} = 25.0$$

$$\text{MSE ("No change")} = \frac{150}{4} = 37.5$$

$$U^2 = \frac{25}{37.5} = 0.67$$

Since the inequality coefficient is less than 1, we can conclude that the current method is producing better results than a forecast of "No change."

5. No single forecasting method consistently gives good results unless conditions are fairly stable. In selecting a method, various techniques should be tried on known data to determine which gives the best results. In addition, the forecasting method actually in use should be tracked constantly to determine when the moment has arrived to change techniques.

9

Inventory Management: Control

CHAPTER OBJECTIVES

Upon completion of this chapter, you should be able to:

A. Use the ABC method to determine the type of control to be applied to items of inventory.

B. Calculate safety stock levels for items with large and small demand.

C. Understand the probabilistic nature of achieving specific levels of customer service.

D. Apply the procedures for managing inventory with both large, regular demand and small, irregular demand.

E. Appreciate the relationship between inventory management and cost reduction.

Applying an equal control effort to all items of inventory would be a great mistake, since not all items are of equal importance. In consequence, the control effort should be in proportion to some measure of importance or value. A variation on Vilfredo Pareto's 80/20 rule tells us that only about 20 percent of the items will require 80 percent of the inventory-control effort. The ABC system of classifying inventory can be used for determining the degree of control needed.

THE ABC METHOD

The ABC method divides the inventory into three (or more, if necessary) categories as follows:

A: high-value items requiring close control.

B: items of medium value requiring only a moderate control effort.

C: low-value items requiring the least control.

Other criteria, of course, can be used in defining each of the three categories (e.g., size, hazardous nature, or frequency of demand), but we shall use cost for purposes of illustration. This classification scheme can be applied to inventory in the following way:

Category	*Value as a Percent of Total*	*Percent of Total Items*	*Type of Records*	*Priority*	*Ordering Procedures*
A	75%–80%	15%–20%	Highly accurate	High	Most accurate
B	15%	30%–40%	Normal	Middle	Use EOQ
C	5%–10%	40%–50%	Simplest	Low	Order at long, fixed intervals

The preceding classification is interpreted as follows:

- Category A items of inventory will comprise 75 to 80 percent of the total value and 15 to 20 percent of the total number of items. Inventory records should be exactingly accurate, with constant monitoring of inventory levels advisable. These items will have the highest priority for the application of management effort.
- Category B items of inventory will comprise 15 percent of the total value and 30 to 40 percent of the total number of items. The company's normal inventory record procedures are used, and reordering is done by

EOQ. These items will have the second priority for the application of management effort.

- Category C items of inventory will comprise 5 to 10 percent of the total value and 40 to 50 percent of the total number of items. Inventory records use the simplest available procedures, and reordering is done at long, fixed intervals. These items will have the lowest priority for the application of management effort.

The following is an example of the way annual cost can be used to select inventory items for each category. Suppose a company wished to classify the ten parts listed in the following table into three categories using the following criteria:

- Category A: The top 20 percent of the items by annual cost.
- Category B: The next 30 percent of the items by annual cost.
- Category C: The remaining 50 percent of the items by annual cost.

According to these criteria, the two items representing the highest annual costs will be placed in category A; the next three items will be placed in category B; and the remaining five items, representing the lowest annual cost, will be placed in category C, as follows:

Part Number	*Annual Usage*	*Unit Cost*	*Annual Cost*	*Category*
1621	50	$0.04	$ 2.00	C
1430	825	0.02	16.50	C
1101	1,275	0.25	318.75	A
1247	115	0.11	12.65	C
1563	1,425	0.12	171.00	B
1415	775	0.17	131.75	B
1209	225	0.01	2.25	C
1320	375	0.15	56.25	B
1057	1,650	0.20	330.00	A
1540	100	0.03	3.00	C

Other classification schemes can be used with the ABC method. If the dollar volume is used, the results are (using the criteria developed previously for assigning items to categories) as follows:

- The two parts listed in category A account for about 62.1 percent of the total dollar volume (i.e., total annual cost of the two items divided by the total annual cost of all items).

- The three parts listed in category B account for about 34 percent of the total dollar volume, calculated as follows:

Part 1563	$171.00
Part 1415	131.75
Part 1320	56.25
Part total	$359.00

$$\frac{\text{Part total}}{\text{Grand total}} = \frac{\$359.00}{\$1{,}044.15}$$

$$= 34.4\%$$

- The five parts listed in category C account for the remaining 3.5 percent of the total dollar volume.

Deciding on how much management effort is to be applied to the various categories of inventory solves an important problem of economizing on this effort. Since not every item is of such importance that each product deserves the same amount of attention, the ABC method of classification adjusts the control effort to the importance of the item. This, in effect, determines what is to be managed. How much stock should be in the inventory is the next problem to be addressed.

DETERMINING STOCK LEVELS

The stock level, or amount of inventory to be carried, can be divided into three large groups. A *safety* stock is held as the last defense against stockouts, a *basic* stock is used during the lead time, or the time it takes to reorder, and a *working* stock is used to fill orders. A reorder would generally be transmitted to a supplier when the working stock has been exhausted and order pickers in the warehouse or distribution center begin using the basic stock. The quantity reordered should then replenish any of the safety stock that has been used and the basic stock and provide a contribution to the working stock.

Since stockouts can be prevented only by maintaining prohibitively large inventories, some method must be used that economizes on the amount of inventory carried but still gives a reasonable degree of protection against stockouts. The safety level of stock provides this protection. The safety stock (sometimes called *reserve* stock) furnishes the items to be used before a stockout occurs and, ideally, will seldom be penetrated except in cases of unexpectedly heavy demand.

Safety Stock When Demand Is Large and Regular

In a more technical sense, the safety stock level represents the amount of variation the inventory can absorb without stocking out. Since absolute protection against all stockouts is impossible, what degree of protection does management want? How much of a risk of a stockout is management willing to take?

A certain amount of mathematical manipulation is required to meet any criteria set by management. The amount of safety stock can be calculated with the following equation:

$$\text{Safety stock} = K \times \sigma$$

where K = factor determined by management policy

σ = standard deviation of demand

Since the calculation of the standard deviation is the more difficult of the two elements in the equation, let's address that first with a definition:

> STANDARD DEVIATION: a measure of the amount of dispersion in the data around some central tendency (usually the mean, or arithmetic average).

Mathematically, in this case we are dealing with what is called a *normal distribution*; that is, all data elements are equally distributed above and below the mean. This produces the familiar symmetrical bell-shaped curve. A large standard deviation describes a low, wide bell curve; a small standard deviation describes a narrow, high-peaked bell curve.

The calculation equation looks formidable but, when translated, boils down to rather simple arithmetic:

$$\sigma = \sqrt{\frac{\sum(X - \overline{X})^2}{n}}$$

where σ = standard deviation of demand

X = each individual demand

$\overline{X}$ = average demand

n = number of individual demands

Let's work through a calculation of a standard deviation using the following demand data for five consecutive months as the initial step in calculating safety stock:

Month	*X* *Demand*	$X - \bar{X}$	$(X - \bar{X})^2$
January	125	$125 - 123 = 2$	4
February	115	$115 - 123 = -8$	64
March	120	$120 - 123 = -3$	9
April	130	$130 - 123 = 7$	49
May	120	$120 - 123 = -3$	9
Total	615	Total	130

$\bar{X} = 615 \div 5 = 123$
$(X - \bar{X})^2 = 130$
$n = 5$

Armed with the information our calculations have produced, the standard deviation can be calculated by substituting the numbers in the equation:

$$\sigma = \sqrt{\frac{\Sigma(X - \bar{X})^2}{n}} = \sqrt{\frac{130}{5}} = \sqrt{26}$$
$$= 5.1$$

Suppose management wants a 99 percent assurance that 95 percent of all orders will be filled without delay (i.e., no more than 5 percent stockouts). The next step is to calculate the safety stock using the standard deviation just computed.

The K factor can be found by using Table 9-1 as follows:

- The 99 percent assurance management wants is γ in the table.
- The 95 percent order fulfillment is P in the table.
- n could be the five demands used to calculate the standard deviation, although n can also be determined by the actual number of demands experienced rather than just those used to calculate the standard deviation.

In using Table 9-1, first follow down the left-hand column to find the 5 used as n. Next locate the column for 99 percent assurance and then the subcolumn for 95 percent order fulfillment. In this $P = 95\%$ column, find the number 7.855 opposite the five demands. We now have

TABLE 9-1
K Factors

Sample size (n)	$\gamma = 0.95$			$\gamma = 0.99$		
	$P = 0.90$	$P = 0.95$	$P = 0.99$	$P = 0.90$	$P = 0.95$	$P = 0.99$
2	32.019	37.674	48.430	160.193	188.491	242.300
3	8.380	9.916	12.861	18.930	22.401	29.055
4	5.369	6.370	8.299	9.398	11.150	14.527
5	4.275	5.079	6.634	6.612	7.855	10.260
6	3.712	4.414	5.775	5.337	6.345	8.301
7	3.369	4.007	5.248	4.613	5.488	7.187
8	3.136	3.732	4.891	4.147	4.936	6.468
9	2.967	3.532	4.631	3.822	4.550	5.966
10	2.839	3.379	4.433	3.582	4.265	5.594
11	2.737	3.259	4.277	3.397	4.045	5.308
12	2.655	3.162	4.150	3.250	3.870	5.079
13	2.587	3.081	4.044	3.130	3.727	4.893
14	2.529	3.012	3.955	3.029	3.608	4.737
15	2.480	2.954	3.878	2.945	3.507	4.605
16	2.437	2.903	3.812	2.872	3.421	4.492
17	2.400	2.858	3.754	2.808	3.345	4.393
18	2.366	2.819	3.702	2.753	3.279	4.307
19	2.337	2.784	3.656	2.703	3.221	4.230
20	2.310	2.752	3.615	2.659	3.168	4.161
50	1.996	2.379	3.126	2.162	2.576	3.385
100	1.874	2.233	2.934	1.977	2.355	3.096

Note: The symbols in this table signify the following: "γ" is the percent of assurance managers want that the computed tolerance will include the desired proportion "*P*" of the prescribed performance standard. The tolerance is calculated by multiplying "*s*", the square root of the standard deviation of demand, by the *K*-factor from this table.

both the standard deviation and the *K* factor for the safety stock computation:

$$\text{Safety stock} = K \times \sigma = 7.855 \times 5.1 = 40.06 \text{ units}$$

If the items cannot be divided into fractions of a unit, we can conclude that the safety stock will consist of 40 or 41 units.

Remember that a few paragraphs back we defined the standard deviation as a measure of the dispersion of data around a central tendency of some sort. If demand is highly variable, we should expect a large standard deviation; if demand is fairly steady, we should expect a small standard deviation. Extending this to a logical conclusion, large demand variability and a large standard deviation result in a larger safety stock; small demand variability and a small standard deviation result in a smaller safety stock.

Safety Stock When Demand Is Small and Irregular

If demand is small and irregular, other procedures must be used to calculate safety stock accurately. Let's examine two of these methods.

One of these techniques uses the service factor (or *f* factor) found in Table 9-2. Management must specify the minimum percent of demand to be met. The maximum number of stockouts permitted can be calculated from this managerial criterion by simply subtracting the demand criterion from 100 percent. The minimum percent of demand to be met (or the minimum percent of stockouts permitted) can then be used to find the appropriate *f* factor from the table.

The level of safety stock that will meet management's criterion of customer service can be calculated using the following equation:

$$\text{Safety stock} = \mu(f\sqrt{a})$$

where μ = average number of units per order

f = service factor from Table 9-2

a = average number of orders arriving during the lead time

An example will be helpful in illustrating the use of this technique. A company makes and sells a product whose customers submit fairly small orders at irregular intervals. The average order is for 15 units, with 10 orders usually arriving during the lead time. Management desires a 90 percent level of service (i.e., wants no more than 10 percent stockouts). Calculation of the safety stock can then proceed as follows:

$$\begin{aligned}\text{Safety stock} &= \mu(f\sqrt{a})\\ &= 15(1.28\sqrt{10})\\ &= 60.7 \text{ units}\end{aligned}$$

Our calculations now tell us that the safety stock should be 60 or 61 units.

The other technique to be illustrated approaches the calculation of the safety stock from the opposite direction: calculating the reorder

TABLE 9-2
f Factors

Minimum Percent of Demand to Be Met	*Minimum Percent of Stockouts Permitted*	*f-Factor*
50	50	0.00
75	25	0.67
80	20	0.84
85	15	1.04
90	10	1.28
95	5	1.65
96	4	1.75
97	3	1.88
98	2	2.05
99	1	2.33
99.5	0.5	2.57
99.6	0.4	2.65
99.7	0.3	2.75
99.8	0.2	2.88
99.9	0.1	3.09

point first and then working back to the safety stock. This method requires the use of a table of Poisson distribution, a portion of which is contained in Table 9-3. If you don't know what a Poisson distribution is, refer to the Mathematical Note at the end of this chapter, but this knowledge is not required to use the technique.

The reorder point is found by entering Table 9-3 with two elements of information: the average demand during the lead time and the level of customer service prescribed by management (e.g., the allowable percentage of stockouts). The average demand during the lead time is calculated as follows:

$$M = D \times L$$

where M = average demand during lead time

D = average daily demand

L = lead time in days

Suppose management specified a 2 percent service level (i.e., no more than 2 percent stockouts), the average daily demand was 3 units, and a

TABLE 9-3
Cumulative Poisson Distribution

Reorder Point (B)	*Stockout probability for average lead time demand ($\overline{M}$)*								
	2	*3*	*4*	*5*	*6*	*7*	*8*	*9*	*10*
2	0.323								
3	0.143	0.353							
4	0.053	0.185	0.371						
5	0.017	0.084	0.215	0.384					
6	0.004	0.033	0.111	0.238	0.394				
7	0.001	0.012	0.051	0.133	0.256	0.401			
8		0.004	0.021	0.068	0.153	0.271	0.407		
9		0.001	0.008	0.032	0.084	0.169	0.283	0.413	
10			0.003	0.014	0.043	0.098	0.184	0.294	0.417
11			0.001	0.005	0.020	0.053	0.112	0.197	0.303
12				0.002	0.009	0.027	0.064	0.124	0.208
13				0.001	0.004	0.013	0.034	0.074	0.135
14					0.001	0.006	0.018	0.041	0.083
15						0.002	0.008	0.002	0.049
16						0.001	0.004	0.011	0.027
17							0.002	0.005	0.014
18							0.001	0.002	0.007
19								0.001	0.003
20								0.002	0.012

Reorder point: $B = \overline{D}L + S$
Average lead time demand: $M = \overline{D}L$
Safety stock: $S = B - \overline{D}L$
Average daily demand: $\overline{D}$
Lead time (in days): L

lead time of 2 days was required to place a replenishment order with a vendor and receive delivery. Average demand during lead time is

$$
\begin{aligned}
M &= D \times L \\
&= 3 \text{ units} \times 2 \text{ days} \\
&= 6 \text{ units}
\end{aligned}
$$

Follow along the row at the top of Table 9-3 until the column is found headed with the number of units in the average lead time demand, in this case the numeral 6. Now go down the column headed by the 6 to find the fraction closest to the designated level of customer service. Fortunately, this time the table contains the precise level of customer service:

0.020. Directly to the left of this 2 percent value, under the heading in the column at the far left of the table, we find the reorder point B of 11 units.

The rest now follows with inexorable logic. The reorder point is composed of the average lead time demand and the safety stock, as follows:

$$B = DL + S$$

where B = reorder point

D = average daily demand

L = lead time in days

S = safety stock

Since $M = D \times L$, we could rewrite this equation more simply as

$$B = M + S$$

and then substitute the numbers we already have as follows:

$$B = M + S$$

$$11 \text{ units} = 6 + S$$

And, by a bit of elementary algebra,

$$S = B - M$$

$$5 = 11 - 6$$

Our calculations now tell us that the safety stock should be about 5 units.

DETERMINING BASIC STOCK

An amount of inventory items (i.e., the basic stock) must be added to the safety stock to determine when a replenishment order should be transmitted to the supplier. As in calculating safety stock, the procedure will differ according to the size of the orders. To keep the two situations differentiated, this basic stock will be termed a *reorder quantity* (ROQ) when demands are small and *economic order quantity* (EOQ) when demands are large.

Inventory Management with Small Orders

The following equation can be used to calculate the reorder quantity (ROQ):

$$\text{ROQ} = \mu \times a$$

where μ = average number of units per order

a = average number of orders arriving during the lead time

Note that μ and a are the same as those used in calculating the safety stock when demands are small. Using the same average of 15 units per order with 10 orders arriving during the lead time, the ROQ is

$$\begin{aligned}\text{ROQ} &= \mu \times a \\ &= 15 \times 10 \\ &= 150 \text{ units}\end{aligned}$$

The reorder point (ROP) is that level of inventory that will generate a replenishment order. The ROP is simply the sum of the safety stock and the ROQ as follows:

$$\begin{aligned}\text{ROP} &= \text{safety stock} + \text{ROQ} \\ &= \mu(f\sqrt{a}) + (\mu \times a) \\ &= 15(1.28\sqrt{10}) + (15 \times 10) \\ &= 60.7 + 150 \\ &= 210.7 \text{ units}\end{aligned}$$

When inventory stocks are reduced to 210 or 211 units, a replenishment order should be prepared equal to the ROQ of 150 units. By the time the replenishment requisition arrives, not all the basic stock will have been issued. Of course, if demand is unusually heavy, the basic stock may be exhausted and part of the safety stock as well. In the latter case, perhaps a sensible course of action would be to order both an ROQ and a safety stock. If, however, some of the basic stock still remains at the time the reorder is received, the excess over the basic stock will be the working stock. Exhaustion of the working stock brings the inventory once again to a reorder point, and the process begins anew.

Managing Inventory with Large Orders

When orders from customers are fairly large, a different ordering system must be used. The reorder point will consist of three elements of inventory:

- A safety stock, which can be calculated using the standard deviation of demand and the K factor in the previous equation: safety stock $= K \times \sigma$.
- Average demand during lead time calculated by the equation just used: $M = D \times L$.
- A reorder quantity called the economic order quantity (EOQ) to distinguish it from the ROQ used when demand is small.

The economic order quantity attempts to minimize two sets of costs: inventory carrying costs (which depend on the number of units in the order), and the administrative costs incurred in placing the order. This problem in cost minimization is portrayed visually in Figure 9-1. The EOQ thus adjusts the carrying costs and the order-processing costs so that total costs are the lowest.

The EOQ can be expressed either in dollars or quantities of units. The following equation is used when calculating the EOQ in terms of dollars:

$$\text{EOQ} = \sqrt{\frac{2ab}{I}}$$

where EOQ = economic order quantity

a = annual usage in terms of dollars

b = administrative costs per order

I = inventory carrying costs per item per year, as a percent of the unit price

Suppose a company used 11,500 units of a particular product per year at a unit price of \$8. The administrative costs connected with placing an order are \$20, while the inventory carrying costs are 20 percent of the price of the item. Translating this information into the symbols used in the EOQ equation yields the following:

$a = 11{,}500 \text{ units} \times \$8 = \$92{,}000$

$b = \$20$ per order

$I = 0.20$

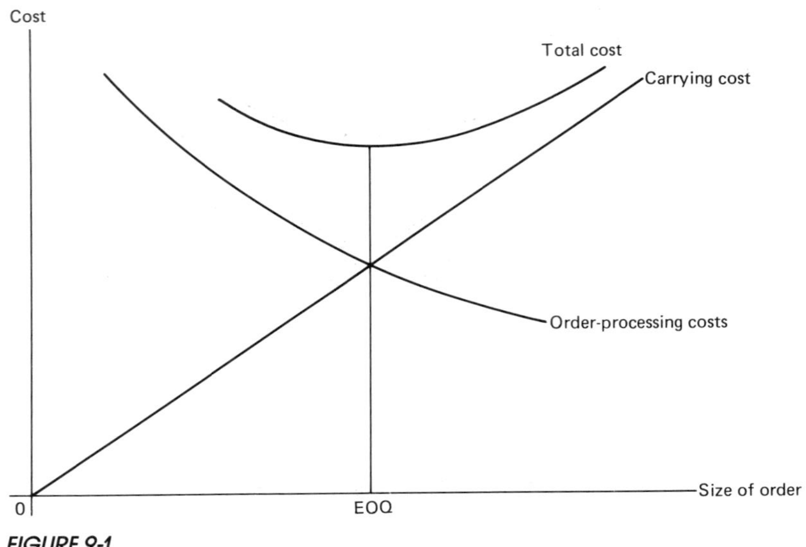

FIGURE 9-1
Determining EOQ

To find the EOQ in terms of dollars per order, substitute this information into the calculating equation as follows:

$$\text{EOQ} = \sqrt{\frac{2(\$92{,}000)(\$20)}{0.20}} = \sqrt{\frac{3{,}680{,}000}{0.20}} = \sqrt{18{,}400{,}000}$$

$$= \$4{,}289.52 \text{ per order}$$

The following equation is used when calculating the EOQ in terms of the physical numbers of units of the product:

$$\text{EOQ} = \sqrt{\frac{2ub}{PI}}$$

where u = annual usage in units

b = administrative costs per order

P = unit price of the item

I = inventory carrying costs per item, as a percent of the unit cost

If we use the same information as in the previous EOQ calculation, we can substitute in the new equation as follows:

$$\text{EOQ} = \sqrt{\frac{2(11{,}500)(\$20)}{\$8 \times 0.20}} = \sqrt{\frac{460{,}000}{1.6}} = \sqrt{287{,}500}$$

$$= 536.19 \text{ units per order}$$

If the two methods of calculating the EOQ are truly equivalent, then multiplying the number of units in an EOQ by the unit price should yield the EOQ in terms of dollars:

$$\text{EOQ (in dollars)} = 536.19 \text{ units} \times \$8$$
$$= \$4{,}289.52$$

And so we have demonstrated that the two methods of calculating EOQ give consistent results.

Let's use this concept of the EOQ in controlling an inventory of a product. Suppose a company has calculated the EOQ to be 120 units. Demand tends to be quite variable, so the safety stock is set at 60 units. Experience has shown the lead time in placing and receiving orders for replenishment purposes to be two days. The first day, the inventory-control worksheet might look like Figure 9-2. Activity for the remaining eight days would be reflected by the entries on the inventory control sheet in Figure 9-2b, as follows:

- Day 2 starts out with an inventory of 150 items left from day 1. Filling demands for a total of 30 items during the day presents no inventory problems.
- On day 3, another demand of 30 items draws down the inventory to 90 items. If demand remains the same on day 4 or even increases, the safety level will be reached. Better, then, to submit an EOQ, which can be delivered on day 5.
- Sure enough, on day 4 a demand of 30 items brings the inventory down to the safety stock. Fortunately, an EOQ was ordered yesterday and is due in tomorrow on day 5.
- During day 5, the EOQ of 120 items arrives on schedule, but another demand for 30 items draws the inventory down to 150 items.
- On day 6 a problem develops: demand doubles to 60 items and draws the inventory down to 90 items. A continuation of this trend in demand will breach the safety level, so an EOQ reorder is placed, which will arrive on day 8.
- Day 7 is a time for mild apprehension. The beginning inventory of 90 items is further reduced to a mere 30 items, warranting placement of another EOQ reorder.
- Like the cavalry to the rescue, an EOQ arrives on day 8, and the day ends with 90 items in stock.
- Day 9 opens with the arrival of the EOQ ordered on day 7. Demand for the day drops back to 30 items, and day 9 ends with 180 items in stock with which to face the uncertain future.

INVENTORY CONTROL WORKSHEET

Item	Day								
	1	2	3	4	5	6	7	8	9
On hand at beginning of day	60								
Due in	120								
Total on hand	180								
Demand	30								
On hand at end of day	150								
Reorder EOQ	–								

(a)

INVENTORY CONTROL WORKSHEET

Item	Day								
	1	2	3	4	5	6	7	8	9
On hand at beginning of day	60	150	120	90	60	150	90	30	90
Due in	120	0	0	0	120	0	0	120	120
Total on hand	180	150	120	90	180	150	90	150	210
Demand	30	30	30	30	30	60	60	60	30
On hand at end of day	150	120	90	60	150	90	30	90	180
Reorder EOQ	–	–	120	–	–	120	120	–	–

(b)

FIGURE 9-2
(a) Inventory control worksheet; (b) using an inventory control worksheet

INVENTORY MANAGEMENT SYSTEMS

Many different inventory management systems can be used when the orders from customers are fairly large. We will examine here four of the major types of systems, which use combinations of safety levels, reorder points, and EOQs to manage the inventory.

Reorder Point/EOQ System

The reorder point/EOQ system (also called the order-point/order-quantity system) will be examined in some detail since many of the elements of this system also apply to the other three to be considered here. This system uses three elements:

- A safety level of stock computed using the previously discussed equation

$$\text{Safety stock} = K \times \sigma$$

- The average demand during the lead time computed using the previously discussed equation

$$\text{Average demand during lead time} = D \times L$$

which, when added to the safety stock yields the reorder point, or ROP:

$$\text{ROP} = (K \times \sigma) + (D \times L)$$

Or, more simply and directly,

$$\text{ROP} = \text{safety stock} + \text{average demand during lead time}$$

- The economic order quantity, or EOQ, computed using the previously discussed equation

$$\text{EOQ} = \sqrt{\frac{2ab}{I}}$$

when an EOQ expressed in dollars is wanted, or

$$\text{EOQ} = \sqrt{\frac{2ub}{I}}$$

when an EOQ expressed in units is wanted.

Inventory is drawn down until the ROP is reached, at which time an EOQ is ordered from the source of supply. The average demand during lead time provides a buffer stock to be used until the replenishment order is filled. Depending on the rapidity of use, this may or may not prevent drawing on the safety stock and thus may not always prevent stockouts. Keep in mind, however, that the level of safety stock has been computed taking into consideration the allowable percentage of stockouts management will accept, so this unfortunate situation should be fairly rare.

This system requires constant monitoring of the inventory to detect promptly when the ROP is being approached. Unfortunately, this system is not well adapted to organizations carrying inventories that contain a large number of different items. Companies producing highly diverse multiple lines would probably be better advised to use some sort of materials requirements planning coupled with a just-in-time (or Kanban) method of control.

The use of the EOQ in placing replenishment orders on suppliers will result in stock levels that vary from period to period and at differing levels above the ROP. For example, notice the situation at time t_2 in Figure 9-3. By the time the EOQ arrives at time t_2', demand has drawn the inventory down to the point where the EOQ barely brings the stock level above the reorder point. After that point, demand becomes so light that the EOQ delivered at time t_3' brings the level of inventory well above the ROP almost to the point of overstockage. The situation would be even more complicated had we not made the rather heroic assumption that the lead time was fixed. A variable lead time is not impossible to handle; it just makes the control more difficult.

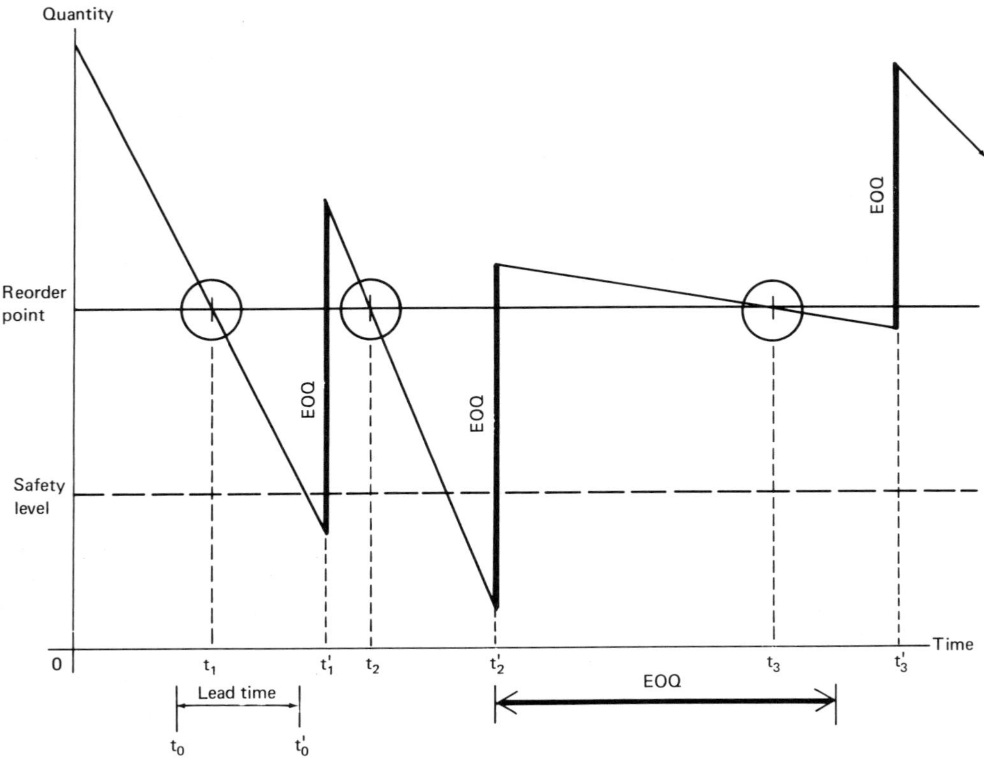

FIGURE 9-3
Reorder point/EOQ method of inventory management

Reorder Point/Maximum Level System

The reorder point/maximum level system (also often called order point/order-up-to-maximum system) uses three somewhat different elements:

- A maximum level of inventory to be maintained as prescribed by management.
- A reorder point (ROP) computed as above by adding the safety stock and the average demand during the lead time.
- A reorder amount.

This system also requires close monitoring of the inventory to detect when the ROP is being approached. When the ROP is reached, an amount is ordered from the supplier to bring the total level of stock up to the prescribed maximum. Although the lead time may be fixed, as in Figure 9-4, the quantity of items in the replenishment order should only be enough to bring the inventory up to the maximum level.

This method requires some skill at forecasting since the demand

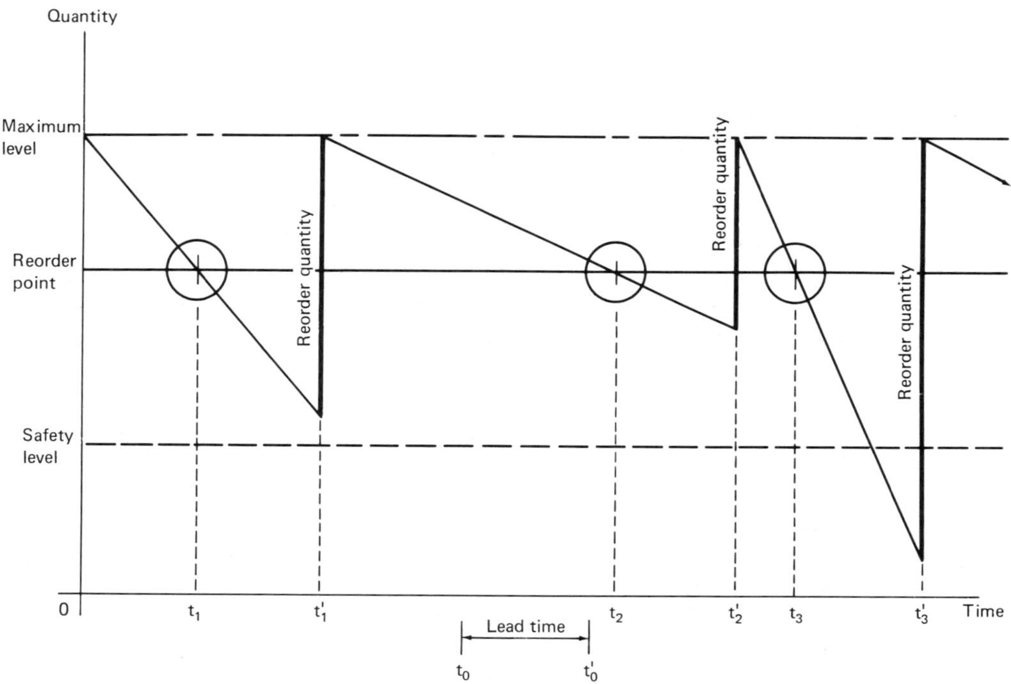

FIGURE 9-4
Reorder point/order-up-to-maximum level method of inventory management

during the lead time may not be known with certainty. The logistics manager will have to estimate how much should be ordered to cover the lead time demand to bring the inventory up to the desired maximum level. Since forecasts often go awry, the maximum level could be either exceeded or fall a bit short, depending on the extent of the forecast errors. The highly variable amounts in the reorders may also cause some problems with suppliers. Vendors are occasionally only able to supply materials in fixed-lot sizes, which may rarely correspond with the quantities desired by the purchaser. This situation will make even more difficult hitting the prescribed maximum level precisely.

The preceding two methods of inventory management require placing replenishment orders whenever a specific event occurs, in this case breaching the reorder point level of inventory. The next two methods require a review of inventory stocks at prescribed intervals of time.

Scheduled Review/Maximum Level System

The scheduled review/maximum level system (also frequently called the periodic review/order-up-to-maximum system) uses only two elements:

- A maximum level of inventory to be maintained as prescribed by management.
- A reorder amount.

This system schedules a review at prescribed intervals of time. At the scheduled review time, no matter what the level of inventory happens to be, a reorder is placed to bring the stock up to the prescribed maximum level, as depicted in Figure 9-5. The safety stock should prevent most stockouts between the scheduled reviews.

Some common sense should be used here. Demand so heavy, if unanticipated, that the safety level is breached should announce an emergency situation even between the review periods. A minor penetration of the safety level, as pictured in Figure 9-5 just before review period t_3, should not set off this early warning system and can be tolerated until the reorder is placed.

The variability in the size of the replenishment orders will also cause some problems. Adjusting the size of the reorder to reach the precise maximum prescribed level will not be possible with any degree of frequency. This same variability will need to be discussed thoroughly with vendors who may prefer to receive orders in fixed sizes. The actual level of inventory, then, can be expected to vary around the maximum level at the time the reorder is received. As a guide to management in

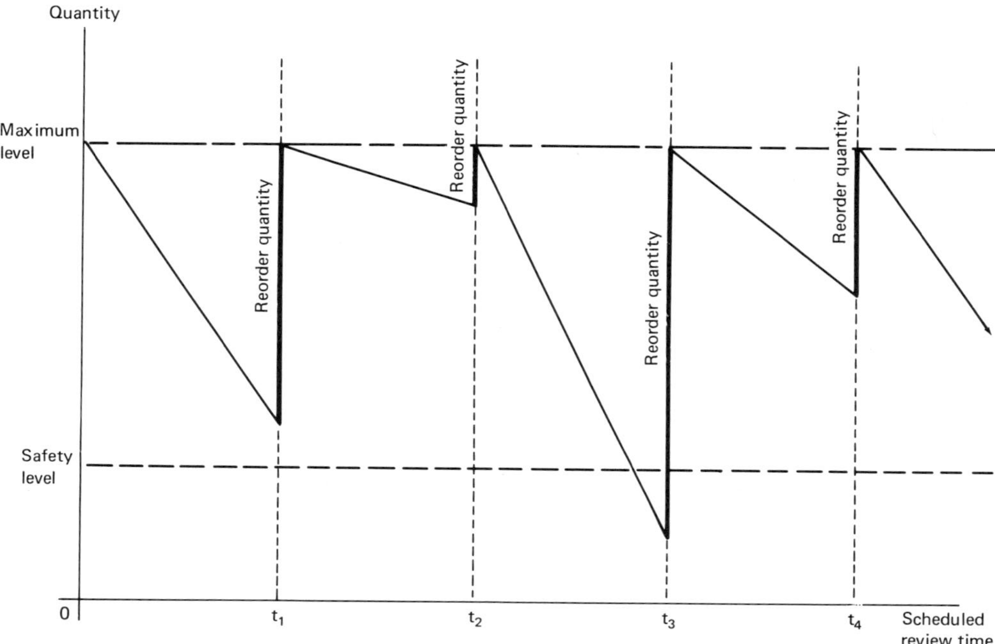

FIGURE 9-5
Scheduled review/order-up-to-maximum level method of inventory management

arriving at a figure for the maximum level, the following equation may be helpful:

$$\text{Maximum level} = K\sigma + \overline{X} + \text{demand during review period}$$

where $K\sigma$ = safety level

$\overline{X}$ = average demand during the lead time

From the foregoing, anyone can understand how much human judgment is involved in managing inventory. Decision rules can be programmed into a computer, but the logistics manager must always be available to react to the unusual, unprogrammed situation.

Scheduled Review/EOQ System

The scheduled review/EOQ system (often called the periodic review/ order quantity system) is perhaps the least satisfactory of the three methods discussed here. Only a single element is used in this system, the

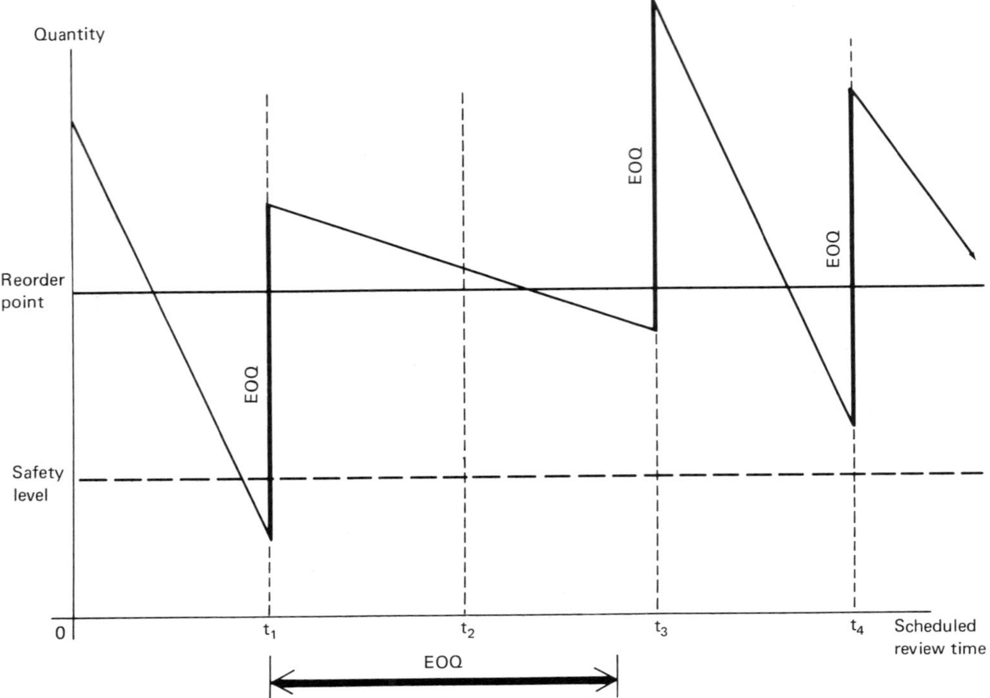

FIGURE 9-6
Scheduled review/EOQ method of inventory management

economic order quantity. The inventory is checked at regularly scheduled intervals, and an EOQ is submitted when the stock is found to be below either the reorder point or safety stock, as in Figure 9-6. This method tends to result in either serious over- or underordering.

Visual Methods of Inventory Control

Two visual methods of determining when to reorder stock are rather easy to use, provided the warehouse personnel can be relied on to give warning when they discover stocks are too low.

The *two-bin method* divides the inventory for each item between two bins or containers. When stock-pickers empty one bin and start on the second, the foreman is notified and a reorder is placed. The *red-line method* requires painting a highly visible red line around the inside of the bin just far enough from the bottom to provide a safety stock. When the stock-pickers empty out the bin to the red line, a reorder is placed. A variation of these visual methods is to package the reserve or safety stock and place the container at the bottom of the bin. When the stock-pickers either reach the package or are forced to open it to fill orders, a reorder is placed.

A corollary to the problem of managing inventory is the task of just keeping track of what is actually on hand. Records, whether maintained by computer or hand, tend to drift out of accuracy. To ensure that the inventory records match the amounts on hand, counting is necessary. Possibly, the worst imaginable way is to shut down the entire firm and for a few days let everyone—clerks, computer operators, machinists, and warehouse personnel—count stock. This method virtually guarantees an inaccurate count. Employees who are not familiar with the stock are assigned to the task of taking a physical count and in many cases they do not appreciate the need for an accurate count. If the company is not completely closed down for the physical inventory and issues continue to be made (or additional stock continues to arrive), the resulting confusion among the untrained inventory takers further degrades the accuracy of the count. With so many people milling around the warehouse, a reconciliation of stock and records becomes impossible with any degree of accuracy.

Cycle counting is a better method. This manner of taking inventory requires only the existing stockroom or warehouse employees, with the possible addition of two or three full-time stock counters if management so desires. Once again using the ABC method of classification, have the trained, experienced stock employees count the A stock (so classified according to dollar value, dollar or physical volume, or whatever criteria are appropriate) once or twice a month, the B stock every three to six months, and the C stock once a year. Each day, a number of items is selected from each category that will meet the prescribed schedule and are counted with as high a degree of accuracy as possible. The counted items are then reconciled with the inventory records, and another set of items is selected for the count on the following day. In this way, the entire inventory gets counted at set intervals and with far greater accuracy than if the warehouse or stockroom is invaded by a horde of counters from all over the company.

Mention should also be made here of the practice of cooperative inventories. Automobile dealers, for example, frequently keep each other informed of their stocks on hand. Should a customer demand from one dealer a model not in stock, the order can be filled quickly by contacting the dealer who does stock the desired car.

One more bit of practical advice. Complementary items (two or more items that are sold and used together) should be ordered at the same time. Thus the wise manager orders frankfurters and buns simultaneously.

LEAD TIME, PRODUCTION, AND INVENTORY COSTS

At this point in the discussion, we should be able to put together some aspects of inventory management to determine when inventories can be eliminated for some items and maintained at a minimum for others. Whenever inventories can be eliminated or reduced, many dollars are being saved for the company while it is still practicing good management. Both over- and underordering can be reduced and production bottlenecks resulting from a shortage of parts or assemblies eliminated by the use of a lead-time control chart.

Every part or subassembly used in the productive process is subject to one or more of the following lead times:

- Lead time to buy from a vendor.
- Lead time to make in the factory.
- Lead time to assemble.

If not located and watched, any one of these lead times can result in a production bottleneck. The lead time in which the company promises to fill orders for customers may also be exceeded if adequate stocks of certain long-lead-time items are not maintained.

The first step in constructing a lead-time control chart is to break the final product, or end item, into its individual assembled subcomponents or parts. Figure 9-7 breaks an end item down in this manner while identifying the lead time. The second step is to construct a lead-time control chart from this breakdown, as illustrated in Figure 9-8. Suppose a company has promised its customers an 8-week lead time in filling and shipping orders. A criterion line is then drawn at 8 weeks horizontally across the chart. An inventory should be maintained on each item crossed by the criterion line. By maintaining stock of these items and keeping them available, the 8-week customer service lead time can be met.

The lead-time chart will also assist management in locating the bottlenecks. Part 11 has a lead time to buy of 16 weeks and is mainly responsible for the 28-week lead time for the finished product. Keeping on hand a stock of part 11 and stocks of the other parts crossed by the criterion line will enable the company to meet its 8-week customer service standard and make more efficient the entire manufacturing operation. We are assuming here that the lead times specified include order preparation. If a company does not include order preparation, then the lead times plotted on the chart will have to be lengthened accordingly.

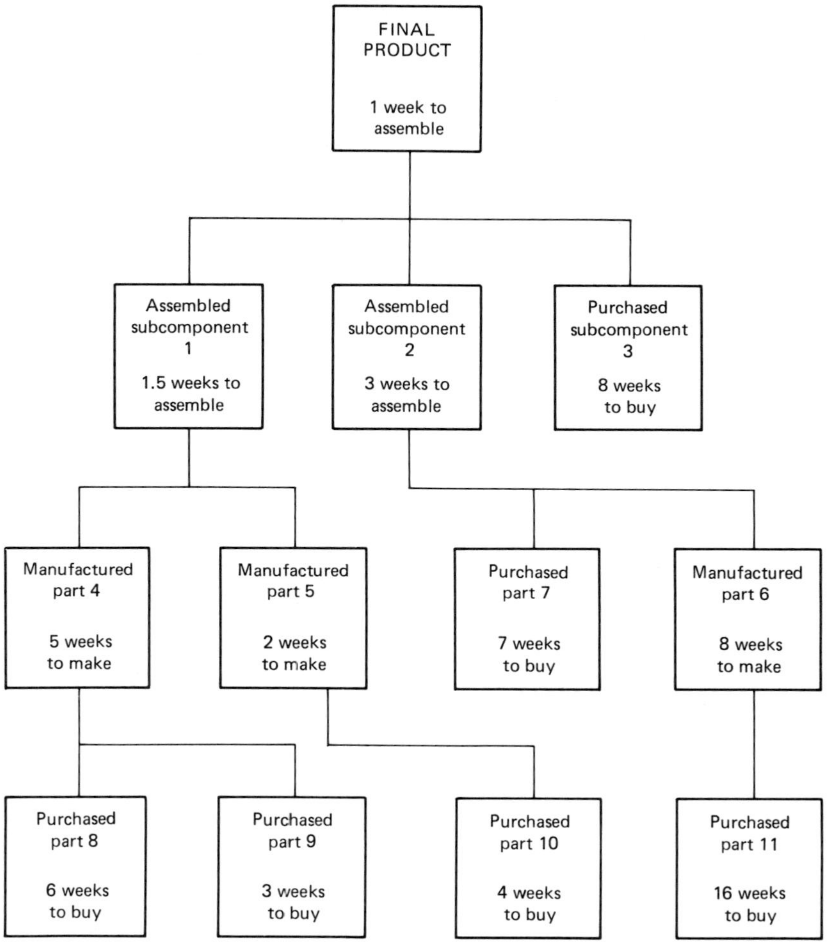

FIGURE 9-7
Lead-time materials chart

If you have now wearied of the domestic scene and are not particularly interested in the mathematical foundations of some of the concepts discussed, we can turn our attention to logistics management in an international setting. Most assuredly, the task gets no easier, for now we will be dealing not just with a variety of organizations, but a variety of national governments as well. When politics enter the situation, complexities abound. Fortunately, as we shall see, plenty of help is available.

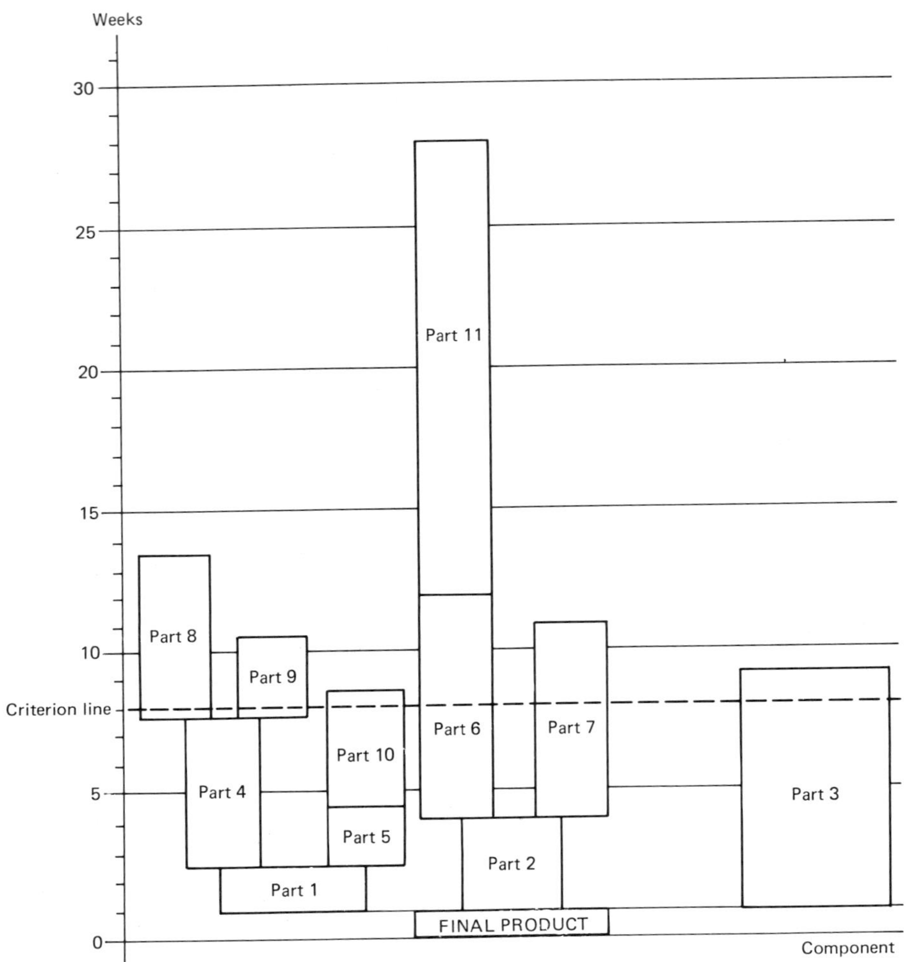

FIGURE 9-8
Item selection chart

MATHEMATICAL NOTES

This section offers an explanation of the mathematical foundations of some of the concepts discussed related to inventory management. None of this is necessary to the continuity of the discussion and can be omitted; but for those with unsatisfied curiosity about the bases for these ideas, the following mathematical notes are offered.

Forecasting

Semiaveraging, or trend analysis, was described as a quick substitute for linear regression. Here's why. Linear regression fits a straight line through the center of gravity of a set of points representing plotted data on the chart. If the set of plotted points appears as in Figure 9-9, fitting this line by eye could result in some serious errors. In fact, either line *AB* or *WZ* would appear to extend through the center of gravity of the points.

Least-squares linear regression seeks to find the proper line by measuring the deviations of each point from the regression line. These deviations are then squared (to eliminate the negative values) and added. The line with the smallest sum of the squared deviations (hence, the name "least squares") offers the best fit. A forecast can then be made by starting at a point on the *X* axis, following directly above this point to the regression line, and then directly over to the left of this intersection to the *Y* axis. The value at this point on the *Y* axis is the forecast. Figure 9-10 illustrates this process. If we now translate the geometric line to an algebraic formula, the familiar equation for a straight line emerges as follows:

$$Y = a + bX \quad \text{or} \quad Y = a - bX$$

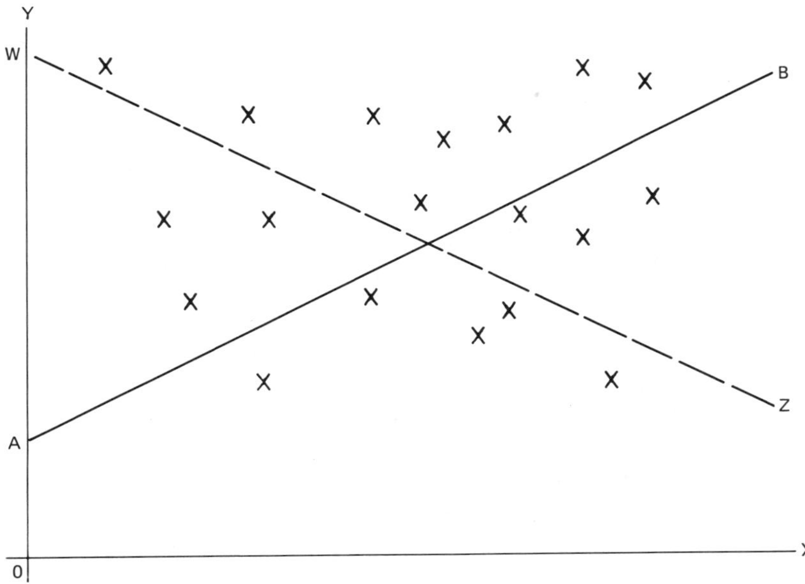

FIGURE 9-9
Possible fitted trend lines

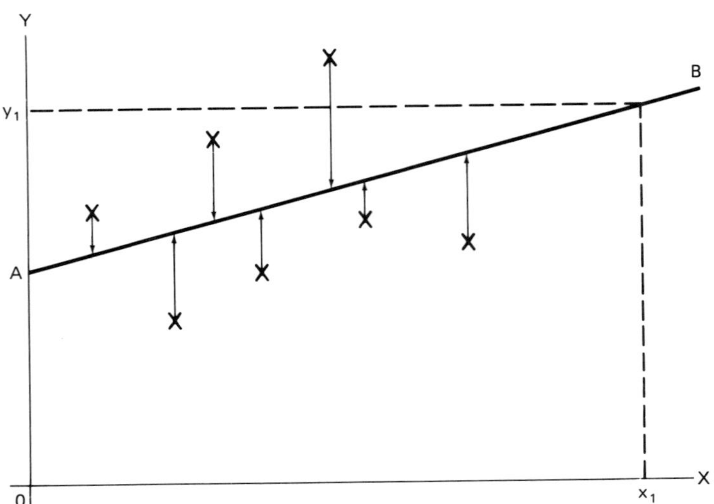

FIGURE 9-10
Line fitted by least-squares regression

where Y = dependent variable being forecast

a = Y intercept, or where the regression line strikes the Y axis (i.e., the value of Y when the X-values are 0)

b = fraction indicating the slope of the regression line; the slope is positive (i.e., upward sloping) when the mathematical sign is plus and negative (i.e., downward sloping) when the mathematical sign is minus

X = independent variable (or the variable that changes first)

The regression equation thus indicates that the forecast Y is the sum of the a value plus a fraction (the b value) of the X data. Any number of random variables (frequently omitted from the equation as we have done) can intervene to introduce errors into the forecast, such as seasonal or cyclical variations, unpredictable events (e.g., natural disasters or unforeseen economic events like recessions or inflation), and human errors in measuring the data used. The forecast, then, can only be regarded as the best estimate of future conditions available at the time the forecast is made.

Now compare the form of the regression equation with the equation used in semiaveraging:

$$F_t = \overline{Y}_1 + \frac{\overline{Y}_2 - \overline{Y}_1}{x} X$$

Notice the similarities:

- F_t is similar to Y, the dependent variable being forecast.
- $\overline{Y}_1$ is similar to a, the Y intercept.
- $\dfrac{\overline{Y}_2 - \overline{Y}_1}{x}$ is similar to b, the slope of the regression line.
- X is similar to the value of the independent variable measured on the X axis.

If variations in the data set being used are not too extreme, the two methods of forecasting give results that are quite close. In any case, regression analysis, although more computationally involved, gives the more accurate results.

Inventory Management

Two different methods of calculating safety stock were presented, depending on whether demand was large or small. The procedure for use with large demands was based on the concept of the normal distribution. This distribution implies that an equal number of demands will lie above and below the average, or mean. When the data are plotted on a graph, the familiar symmetrical bell-shaped curve emerges, as illustrated in Figure 9-11. Different demand experiences will produce differently shaped bell curves. The only requirement of the normal distribution is that the demand data be distributed equally above and below the mean. Part of the usefulness of the normal distribution is that the median and the mode of the data are at the same point as the mean.

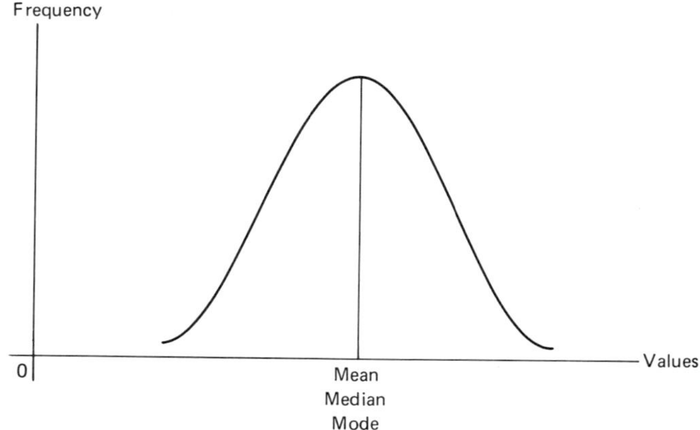

FIGURE 9-11
Normal distribution

To give you an idea of the mathematical complexity of this distribution, here is the equation that describes the normal distribution:

$$y = \frac{n}{\sqrt{2\pi}} e^{-x^2/2\sigma^2}$$

where y = frequency of the X values

x = deviation from the mean of each element in the distribution

n = total frequency, or number, of items in the distribution

σ = standard deviation

e = the Naperian constant 2.71828

π = the constant pi, which equals 3.14159

The normal distribution thus represents the relationship between the frequency of measurements (on the Y axis) and the deviations of these measurements (on the X axis) from the mean.

A situation in which demand might be expected to be large and thus make use of the assumption of the normal distribution valid might occur at the factory level, where fabrication of an item would require a heavy use of subassemblies and components. Suppose, however, we are at the retail level, where demands could easily be small and perhaps irregular in arrival as well. In such a case, the bell-shaped curve described by the distribution may not be symmetrical but canted to either the right or left. The individual data elements would thus not be equally distributed in the usual perfectly symmetrical bell curve. When the demand data are not equally distributed above and below the mean, the Poisson distribution could be used (named after the French statistician who worked it out) to describe the skewness in the data. The equation describing the Poisson distribution is somewhat less complicated:

$$y = e^{-m} + \frac{m^c}{c!}$$

where y = frequency of the X values

$m = Np$ (where N is the size of the sample and p is the probability of occurrence)

c = number of times the event occurs in the sample

As the demands become larger, the Poisson distribution tends to shade into a normal distribution.

Economic Order Quantity

The equation for calculating the economic order quantity (EOQ) is derived as follows: First, set up the equation for total costs (which we are trying to minimize):

$$TC = ac + \frac{ab}{Q} + \frac{QI}{2}$$

where a = annual usage in units

b = administrative cost of placing an order

c = acquisition cost per item

I = carrying cost, which is equal to cost per item expressed as a percentage of the unit acquisition cost

Q = units per order

Second, minimize this expression of total cost using differential calculus. The first derivative of the total cost equation (or function) of quantity with respect to total cost is

$$\frac{dTC}{dQ} = \frac{I}{2} - \frac{ab}{Q^2}$$

Third, set the equation for the first derivative equal to zero and solve for Q^2:

$$\frac{I}{2} - \frac{ab}{Q^2} = 0$$

which by a bit of algebra becomes

$$\frac{I}{2} = \frac{ab}{Q^2}$$

$$\frac{Q^2 I}{2} = ab$$

$$Q^2 I = 2ab$$

$$Q^2 = \frac{2ab}{I}$$

Finally, extract the square root of Q^2, and the equation for the economic order quantity emerges:

$$Q = \sqrt{\frac{2ab}{I}}$$

OBJECTIVES ACHIEVEMENT CHECKUP

The following exercises will help you in assessing the extent to which you have achieved the objectives for this chapter. Where essay-type answers are required, the more you are able to condense your knowledge into a short paragraph of a few sentences the better. Looking back as you work these exercises and rereading parts of the chapter are permissible; but do not look ahead to the answers until you have done your best with the questions.

1. You are in the process of installing a system of cycle counting and wish to classify items of inventory using the ABC method to determine the frequency of the physical count. The following information is available to you:

Part Number	*Annual Usage (units)*	*Unit Cost*
1042	1,000	$0.15
1073	200	1.10
1081	5,000	0.45
1104	7,500	0.25
1111	1,500	1.35
1117	300	3.50
1137	400	0.05
1152	3,500	0.20
1171	8,000	1.00
1178	2,500	0.80

   a. You use the ABC method of classification as follows: The items that account for 80 percent of the total cost are to be placed in class A, the items that account for the next 15 percent of the total cost are to be placed in class B, and the items that account for the remaining 5 percent of the total cost are to be placed in class C. Which items in the preceding list would be placed in each category?

   b. Would your assignment of items to each category change if the basis of the classification was changed to annual usage, with the items with the highest use comprising 20 percent of the total items in class A, the items with the next highest use comprising 30 percent of the total items in class B, and the remaining items in class C?

2. Your research reveals that demand for a particular item is fairly large and regular, as follows:

Month	*Quantity Demanded*
June	115
July	130
August	135
September	125
October	120

Your supervisor tells you that an assurance of 95 percent is wanted that 90 percent of all orders will be filled without delays due to stockouts.

a. How much safety stock will be required to meet these criteria?

b. By how much will the safety stock have to be increased if management wants only a 1 percent level of stockouts?

3. You discover that demand for a particular product is both fairly small and highly irregular. The average order is for only 7 units, with 5 orders usually arriving during the lead time. Management wants no more than 10 percent stockouts.

   a. Calculate the amount of safety stock you would recommend in this situation using the service factor method.

   b. Calculate the amount of safety stock using the Poisson distribution table. Use the following information to find the reorder point:

   Average daily demand: 15 units
   Permissible level of stockouts: 5 percent
   Lead time: 10 days

4. Calculate the reorder quantity (ROQ) when the average number of units per order is 7 items and the average number of orders arriving during the lead time is 5.

5. What will be the reorder point when management specifies no more than 10 percent stockouts? Use the following information in your calculations:

   Average number of units per order: 3
   Average number of orders arriving during the lead time: 5

6. You have the following information available to you as you start to calculate a reorder point:

Month	*Quantity Demanded*
April	1,250
May	1,400
June	1,300
July	1,200
August	1,350

Your supervisor tells you that a 99 percent assurance is wanted that stockouts will not exceed 5 percent. In addition, your records show that the lead time in days is 15 and that the average daily demand is 75 units. Inventory records reveal that the annual use of this particular item is 20,000 items, the administrative cost of placing an order is $15 per order, the price of the item is $30, and the inventory carrying cost as a percent of the unit price is 10 percent. Analyze this situation in terms of safety stock, economic order quantity, and reorder point.

7. Why should cost reduction be an important consideration in inventory management?

ANSWERS TO CHECKUP QUESTIONS

1. The following table contains the answers to both parts a and b.

Part Number	*Total Unit Cost*	*Percent of Cost*	*Classification by:* Cost	*Classification by:* Usage
1042	$ 150	1	C	C
1073	220	1	C	C
1081	2,250	12	A	B
1104	1,875	10	B	A
1111	2,025	11	A	C
1117	1,050	6	B	C
1137	20	Negligible	C	C
1152	700	4	C	B
1171	8,000	44	A	A
1178	2,000	11	A	B
Totals	$18,290	100		

2. The first step is to calculate the standard deviation of the quantities demanded:

Month	*Demand (units)*	$X - \bar{X}$	$(X - \bar{X})^2$
June	115	115 − 125 = −10	100
July	130	130 − 125 = 5	25
August	135	135 − 125 = 10	100
September	125	125 − 125 = 0	0
October	120	120 − 125 = − 5	25
Totals	625		250

$$\sigma = \sqrt{\frac{\Sigma(X - \bar{X})^2}{n}} = \sqrt{\frac{250}{5}} = \sqrt{50}$$

$$= 7.1 \text{ units}$$

a. Safety stock with 90 percent level of customer service. Data needed to find the *K* factor:

$$\gamma = 95\%$$

$$P = 90\%$$

$$n = 5$$

$$K = 4.275$$

Since the standard deviation is 7.1 units, the safety stock will be

$$\text{Safety stock} = K \times \sigma$$

$$= 4.275 \times 7.1$$

$$= 30.4 \text{ units}$$

b. Safety stock with 99 percent level of customer service. Data needed to find the new *K* factor:

$$\text{New } P = 99\%$$

$$\text{New } K = 6.634$$

Using the standard deviation previously calculated of 7.1 units, the safety stock will be

$$\text{Safety stock} = 6.634 \times 7.1$$

$$= 47.1 \text{ units}$$

And the increase will be 47.1 − 30.4 = 16.7 units.

3. $$\text{Safety stock} = \mu(f\sqrt{a})$$

$$= 7(1.28\sqrt{5})$$

$$= 20 \text{ units using the service factor method}$$

The first step in using the new data and the Poisson distribution table is to calculate the average daily demand during the lead time:

$$M = D \times L$$

$$= 15 \text{ units} \times 10 \text{ days}$$

$$= 150 \text{ units}$$

Use the Poisson table to find the reorder point. Using a stockout probability of 0.049 as the closest approximation to the specified 5 percent, the reorder point is 15 units. Since $B = M - S$, the safety stock will be

$$S = B - M$$

Therefore, the safety stock will be

$$\text{Safety stock} = 15 - 150$$

$$= 135 \text{ units}$$

4.
$$\begin{aligned} ROQ &= \mu \times a \\ &= 7 \times 5 \\ &= 35 \text{ units} \end{aligned}$$

5.
$$\begin{aligned} ROP &= \text{safety stock} + ROQ \\ &= \mu(f\sqrt{a}) + (\mu + a) \\ &= 3(1.28\sqrt{5}) + (3 \times 5) \\ &= 8.6 + 15 \\ &= 23.6 \text{ units} \end{aligned}$$

6. The first step is to calculate the standard deviation of demand:

Month	*Demand (units)*	$X - \overline{X}$	$(X - \overline{X})^2$
April	1,250	1,250 − 1,300 = − 50	2,500
May	1,400	1,400 − 1,300 = 100	10,000
June	1,300	1,300 − 1,300 = 0	0
July	1,200	1,200 − 1,300 = −100	10,000
August	1,350	1,350 − 1,300 = 50	2,500
Totals	6,500		25,000

$$\sigma = 70.7 \text{ units}$$

$$\begin{aligned} \text{Safety stock} &= 7.855 \times 70.7 \\ &= 555.3 \text{ units} \end{aligned}$$

The average demand during the lead time is

$$\begin{aligned} M &= D \times L \\ &= 75 \times 15 \\ &= 1{,}125 \text{ units} \end{aligned}$$

And the economic order quantity (EOQ) is

$$EOQ = \sqrt{\frac{2\,ub}{PI}} = \sqrt{\frac{2\,(20{,}000)(\$15)}{\$30 \times 10\%}} = \sqrt{\frac{(40{,}000)(15)}{3}} = \sqrt{\frac{600{,}000}{3}}$$

$$= 447.2 \text{ units}$$

If virtually all the available stock has been depleted or a stockout has actually occurred, it may be best to order both a safety stock, an average demand during lead time, and an EOQ as follows:

Safety stock	555.3 units
Average demand	1,125.0
EOQ	447.2
Total	2,127.5 units

If a stockout has not yet occurred, the reorder point will be the sum of the average demand during the lead time and the safety stock:

Safety stock	555.3 units
Average demand	1,125.3
Total	1,680.3 units

Thus, when the inventory is drawn down to 1,608.3 units, an EOQ of 447.2 units should be placed as a replenishment order.

7. Cost reduction should be an important element in inventory management for the following reasons:

   a. Inventory represents money; any effort to reduce the size of the inventory contributes to cost reduction.

   b. Inventory costs include the costs to acquire the stock, the costs of keeping the stock in storage, and occasionally the cost represented by inventory taxes paid to local jurisdictions.

   c. Excessive inventories represent losses due to shrinkage of the inventory while in storage and the lost opportunities to make better use of the funds tied up.

part 3

BEYOND THE HORIZON IN SPACE AND TIME

10

International Trade and Logistics

CHAPTER OBJECTIVES

Upon completion of this chapter, you should be able to:

A. Know the relationship between logistics management and trade.

B. Identify the differences and similarities between domestic and international logistical functions.

C. Understand the effect of changes in foreign exchange rates on inventory values.

D. Describe the methods of reducing the risk connected with holding inventories abroad.

E. List the functions of firms that specialize in handling international movements of freight.

The next time you are delayed at a traffic intersection by a red light, observe the passing cars and their nations of origin. Turn over the spoon at a restaurant and note the area of manufacture. Cars from Germany and Japan, tableware from Taiwan, shoes from Italy and Brazil, fine fabrics from South Korea—where does it all end? Even in Europe, the traveler in Germany spreads Danish butter on the French croissant in the Gasthaus in Heidelberg. The reverse is also true. We have brought the misery of American dietary habits to all corners of the globe.

How are all these vast flows of goods around the world made possible? By logistics, of course. When the means of logistics were rudimentary and the routes of travel often blocked by wars and hostile nations, trade in medieval Europe was a mere trickle. Even salt was a luxury high in price and short in supply. When logistical facilities improved and expanded, so did trade. Add sails to replace oars in ships and suddenly the North American colonies had sugar and rum to satisfy their palates.

Given modern means of transportation and storage, the list of items moving between nations grows endlessly. For example, between 1970 and 1982, world exports grew by about 484 percent despite a serious worldwide economic recession. Table 10-1 indicates a comparable performance for imports. A glance at the total of imports plus exports for each year in Table 10-1 reveals that we are dealing with an international logistical operation of staggering proportions.

ORGANIZING FOR INTERNATIONAL LOGISTICAL OPERATIONS

Many companies entering international trade seem to edge into such functions gradually. A company finds an economical source of raw materials or semifinished goods and begins importing on a modest scale. If this practice proves profitable, the imported materials are worked up into finished goods for domestic buyers. Perhaps the product eventually finds favor with a foreign buyer, and a reverse flow of goods begins. If the product catches on and sales expand, exports to buyers overseas grow. Soon shipments are made to an overseas regional distribution center rather than to individual buyers. The final step is to buy, build, lease, or franchise an assembly or manufacturing operation in one or more foreign locations. If the company manufactures military goods (e.g., fighter aircraft, missiles), the problem of logistics grows enormously since several nations are often involved in the design, testing, and supply of components, repair parts, and end items.

How do companies respond to these growing problems? Initially, when the volume of shipments is small, a single employee may be

TABLE 10-1
World Trade Between 1970 and 1982

Year	Imports (in millions)	Change (%)	Exports (in millions)	Change (%)
1970	$ 328,597	—	$ 314,309	—
1975	903,980	+175	874,948	+178
1981	2,035,229	+125	1,978,973	+126
1982	1,919,684	−6	1,832,022	−7
	Total change 1970–1982	+484	Total change 1970–1982	+483

assigned to the export function, perhaps in the traffic department. Financial matters, transportation arrangements, preparation of documents are all handled by the people who usually perform these functions for domestic shipments. Eventually, a separate department is organized to deal with the details of export/import movement of goods. That the conduct of an import/export business soon becomes quite beyond the capability of the employees who handle domestic logistical functions should come as no surprise.

DIFFERENCES AND SIMILARITIES

The broad outlines of the functions of domestic and international logistics management are similar. Orders must be obtained and filled, goods must be stored and shipped; and the usual functions of transportation, customer service, and documentation must all be performed. These similarities disappear rapidly when the details are examined more closely.

Some of the differences that set apart domestic and international logistics management are cultural. In many areas of the world goods are wanted only in small quantities. Where adequate storage facilities are lacking, stocks in retail outlets are small, thus forcing consumers to shop at frequent intervals. In the absence of refrigerators or freezers, shopping for perishables is a daily affair. During the war in Vietnam (and perhaps at other times as well) cigarettes were sold on the streets of Saigon by sidewalk merchants not by the carton or pack, but by the individual cigarette. Large stocks of goods cannot be thrust on a retail establishment this small. In such cases, the distribution channels must carry a higher level of supply.

In the United States and Japan, packaging is as much a part of the merchandising effort as other advertising; elsewhere in the world, much retail merchandise is handed over to the consumer wrapped in newspaper. Still the packaging of most goods must be sturdy enough to withstand the strains of long-distance transport, giving marketing and engineering experts the customary problems of solving the trade-off between attractive sales appeal and protection.

Transportation networks often present unusual problems for logisticians and national governments to solve. During the occupation of Korea from 1895 to 1945, the Japanese authorities concentrating on building a north–south rail net to serve as a connecting link between the sea routes to the south and other rail links to the north through Manchuria and Russia. After World War II, the free government of South Korea was faced with the problem of building a new network to serve points on routes running east and west. Capacities of railcars and sizes of track differ, changing at national boundaries, to add to the logistical problem. The American standard gauge track with a width of 4 feet $8\frac{1}{2}$ inches is replaced in India by the broad gauge 5-foot track width. Port handling facilities in some areas are so limited even cargo normally shipped in bulk may have to be packaged. In less developed tropical areas, workers carry bagged sugar (sacked to permit this type of hand loading and unloading) on their shoulders since the port areas lack loading chutes, which would direct a stream of sugar directly into the hold of the vessel.

Perhaps the most difficult difference to handle results from pure nationalism. A bias frequently exists in favor of local merchants, producers, and suppliers, which all but closes the distribution channels to foreign companies. A common reaction is for the foreign producers to ship components overseas for local assembly and marketing. On the other side of the ledger, this type of acquisition of overseas facilities also risks a takeover—with or without compensation—should a local unstable government so desire.

Imagine the problems generated by the need to handle returned items when the rejected or defective products have already moved across several national boundaries with the customs duties already paid! Here is an excellent opportunity for just about everyone in the company, from the logistics manager to the accountant, to indulge in some highly effective interdepartmental cooperation.

HANDLING INVENTORY

Inventories perform the same functions in international logistics as in domestic logistics, with a few extra management problems thrown in. Inventory overseas provides the usual stock of items from which orders are filled. The management of this inventory, however, is complicated overseas by the longer transportation time and the impact this has on the length of the order cycle. Time, in this instance, has several dimensions.

The longer cycle means that the products are in the logistical pipeline for longer periods of time and are handled at more points. This increased time and handling exposes shipments to the multiple risks of changes in laws and regulations governing imports and exports, pilferage en route, deterioration, and obsolescence. Even keeping track of changes in foreign taxes and import–export duties is troublesome when the lines of communication and logistics stretch halfway around the world. Overseas subsidiaries or distributors can ease some of this burden of inventory management, but not all. For example, the risk of fluctuations in foreign exchange rates is a problem peculiar to international logistics that only careful management can minimize.

The financial risk connected with the changing prices of foreign currencies is termed *exposure*:

> EXPOSURE: The risk of gain or loss resulting from holding inventory stock overseas the value of which changes as currency exchange rates fluctuate.

Dealing with changes in the values of inventories due primarily to changes in the prices of national currencies involves offsetting both gains and losses that result from these value changes.

Perhaps an example will best illustrate both the problem and the solution. Suppose the British subsidiary of an American firm provides the following information:

> Value of inventory on hand on January 1, 19AA: \$1,250,000
> Current exchange rate: £1 = \$2.50

With the British pound exchanging for \$2.50 U.S. dollars, the inventory is thus valued in England at £500,000 on the records of the subsidiary.

To the consternation of management, the company's financial experts predict that the pound will continue to weaken in the coming year and will probably end up worth only \$2.40. The parent U.S. company thus faces a *translation loss*:

> TRANSLATION LOSS (or GAIN): The decrease (or increase) in the dollar value of an inventory held overseas due to changes in exchange rates.

Keep in mind what is happening here. The market price of the inventory held in the foreign country will not change. What *does* change is the price of the currency in which the inventory is valued, and this causes the change in the dollar value of the inventory on the records of the parent U.S. company.

If the exchange rate does in fact fall to £1 = $2.40, the company will end the year with an inventory valued in Britain at £500,000 but valued in U.S. currency at $1,200,000 in America. This represents a shrinkage in the value of the overseas inventory, or a translation loss, of $50,000 to the parent company.

The American company will, of course, want to protect itself from this loss if at all possible. This is what could be done to offset the loss: On January 1, 19AA, contract with a speculator in foreign exchange for the delivery of £500,000 of British currency on January 1, 19BB, one year later. This contract, known as a forward contract, in foreign exchange will bring the company $1,250,000 at the current price of £1 = $2.50.

One year later, on January 1, 19BB, the British money must be delivered to the speculator. The parent U.S. corporation now buys the £500,000 promised to the speculator at the current exchange rate of £1 = $2.40, or $1,200,000. The company then hands the £500,000 to the speculator, is paid under the terms of the contract at the contract price of £1 = $2.50, and receives $1,250,000 from the speculator in return for the British currency.

The company has now gained $50,000 on the contract with the speculator by selling the promised £500,000 at the contract price of $2.50 and buying the promised British pound currency at the current market price of $2.40. This gain of $50,000 has been exactly offset by the $50,000 loss on the value of the inventory, with a net result of *zero*.

If this procedure worked perfectly, translation gains and losses would be no problem; but perfection is frequently an ideal, not a reality. If the forecasts of the prices of the two currencies are wrong, either by the speculator or the financial experts of the company, the net effect may not turn out to be zero. As a result of these human imperfections in forecasting foreign exchange rates, the reports of companies engaged in international trade frequently list translation gains and losses on their books.

Other methods of handling inventory risks can be used if the company does not want to engage in currency speculation. Diversification is one available alternative. A company could obtain raw materials and produce in nations whose currencies are not subject to frequent variation. Since the prices of national currencies are now determined by supply and demand on the foreign exchange markets rather than by reference to a given amount of gold, such a national currency might be difficult to find. The price of the Japanese yen, for example, once stable for years at ¥360 = $1, has on more recent occasions gyrated between

¥180 and ¥240 to the U.S. dollar. In foreign trade, though, the criteria of stability and price of the national currency must be applied with common sense. The Laotian kip, worth about 3 cents, does not change very much, but then, how many concerns are eager to establish manufacturing and sales facilities in Laos?

Another method of diversification would be to produce in one country but hold the inventory in a different country. This procedure is analogous to producing within the United States at one location but holding the inventory of finished products at a distant regional distribution center, which might make performance of the distribution function more efficient.

Changing the timing of the order cycle could help reduce the risk of holding inventories abroad. The timing could be reduced to hold inventories to a minimum and use expedited means of transportation to meet delivery requirements. Or, in contrast, the timing could be increased to keep the logistical pipeline full to allow time for the shipment to arrive just at the desired time with a minimum of storage.

A rather delicate balance must be struck by the international logistics manager. Reducing the exposure of the firm to currency price changes may not always provide the lowest-cost logistics management. The decision is, thus, a complex one compounded of the desire to penetrate and remain in selected foreign markets, balance costs against revenues, and give proper consideration at the same time to meeting the needs of customers.

A particularly fruitful way of achieving stability in the valuation of inventory is to use economic unions. The most prominent of these economic unions is the European Economic Community (often and more popularly called the European Common Market), consisting of Belgium, France, West Germany, Italy, Luxembourg, and Holland when first formed in 1958, and to which the United Kingdom, Denmark, Greece, and Ireland were added more recently. Although less well known, other economic unions are scattered throughout the world. The Central American Common Market, established in 1961 to form a common market consisting of Costa Rica, El Salvador, Guatemala, and Nicaragua; the European Free Trade Association, formed in 1960 of nations denied membership in the European Common Market and consisting of Austria, Iceland, Norway, Portugal, Sweden, and Switzerland (with Finland as an associate member); and the Association of Southeast Asian Nations, formed in 1967 to stimulate economic development in the region occupied by Indonesia, Malaya, the Philippines, Singapore, and Thailand are but a few examples.

These economic unions are really huge multinational trading zones in which customs barriers have been reduced or eliminated to promote free trade among the members. Such free-trade zones often permit delaying the payment of duties and other fees until the shipments move out of the zone to a final destination.

SPECIALIZED FACILITIES

Foreign trade involves a company in a number of additional functions. More handling is required as the goods pass through air or water ports, and documentation requirements are greatly increased to include export declarations, various types of invoices, tally sheets used to count the cargo each time the mode of transportation changes, insurance policies, and delivery permits. As if this were not enough, government restrictions on the free movement of goods must be dealt with, such as tariffs, licenses, quotas, embargoes, currency controls, and whatever other revenue-producing restrictions a local government can devise.

Fortunately, much help is also available to meet these problems and complications.

Specialists in Foreign Trade

When the logistics personnel of a company find the conduct of the international affairs of their company more than they can handle comfortably without significantly increasing the size of the staff, resort can be had to those who specialize in various aspects of foreign trade.

The *international freight forwarder* actually does far more than the domestic freight forwarder. This international specialist acts as an agent for the exporting firm in a number of ways. Arrangements are made for the movement of goods to both air and water ports and includes such functions as booking cargo space on board aircraft and ocean vessels, preparing documentation (e.g., government export declarations, bills of lading, and consular documents), obtaining insurance, and, if required, acting as the export department of the shipping company by giving advice on market conditions overseas and taking orders for goods.

The *non-vessel-operating common carrier* (NVOCC) is a unique specialist in foreign trade for which no domestic counterpart exists. Although operating no vessels, this unusual type of common carrier enters into joint rates with ocean carriers (but not with domestic rail or highway carriers who might carry the freight to the port of departure). The NVOCC consolidates various small less-carload shipments into larger shipments that will fill a shipping container. In general, the NVOCC is responsible for getting the merchandise from the shipper's warehouse to the overseas customers.

An *air freight forwarder* performs the same functions for air shipments as the international freight forwarder. The air freight forwarder can be either an independent company specializing in air shipments or part of a larger international surface freight forwarder.

The *customs-house broker* performs freight-forwarding functions for an importer, just as the international freight forwarder performs services for the exporter. This type of broker clears shipments through customs (a bothersome task at best in many foreign ports with the possible sole exception of Bahrein, which encourages trade by charging no import or export duties), pays the various fees, and arranges for the transportation of goods after clearing through the port.

The *export management company* is perhaps in the best position to serve as an extension of the logistical organization of a shipper. The functions performed by these companies are rather wide ranging. They provide assistance to companies in finding firms overseas that are able to manufacture the product, handle overseas sales, provide credit information on prospective buyers in foreign countries, and can link their sales and distribution services with a freight forwarder.

The *export packer* does what the name implies: packs shipments for export. For example, in nations that insist on examining the contents of crates and boxes, the export packer can provide special cartons with slatted sides to permit visual inspection and still protect the cargo.

International Transportation Facilities

If we exclude submarine freight service (which is still not offered commercially), air and water dominate the international transportation scene.

Air transportation is mainly of two types: charter service, in which an entire aircraft is leased, and scheduled service, which uses space in passenger-carrying aircraft. A few all-cargo scheduled airlines also exist.

The use of air transportation greatly accelerates transit time. At the same time, even large all-cargo aircraft can carry only a limited amount of freight. Limitations on carrying capacity, however, are offset to a considerable degree by sheer speed. One all-cargo jet aircraft can carry only about 1 percent of the capacity of a small ocean-going vessel; but the aircraft covers in one hour up to 35 times as much distance as the surface vessel. Transit time can thus be reduced from days or weeks to just hours.

In cases of extreme emergency, even the most unsuitable cargo can be carried by air. For example, when the Russians blocked surface travel to Berlin in 1948–1949, the Allied powers loaded coal into aircraft and flew over the blockade. Other than in such extreme cases, cargo more suited to air transportation is shipped. The containers used in air transport vary considerably to accommodate the packing to the cargo spaces available. When pallets are used, fiberglass igloos are often used to encase and secure the loaded pallets.

Competition in the international air transport industry is regulated in three ways: (1) by local governments who control access to their airports and can impose whatever regulations they see fit; (2) by treaties between governments, which establish mutually agreeable regulations; and (3) by the International Air Transport Association (IATA), which exercises a certain degree of rate control.

The IATA specifies three types of rates. The first is a general rate, which is a basic rate used for all goods when no other type of rate is available. It bears a strong resemblance to the class rates used in domestic transportation. Second, IATA commodity classification tariffs are used to construct the rate for a particular commodity as a specified percentage of the general rate provided for each class. The third type of rate is familiar from domestic transportation: a specific commodity rate established for the movement of a particular commodity between specific points.

Ocean transportation, although slow (a popular song once featured a slow boat to China as the ideal vehicle for romance), has the advantage of cost over air carriage. In fact, the versatility of surface water transportation frequently makes it the only feasible way of hauling many types of products. Four main types of ocean carriers are available to shippers.

Common carriers sail on fixed routes and established schedules. The conventional cargo vessel used by these common water carriers hauls a variety of general cargo; but ships that handle highly specialized types of freight are becoming more common. The container vessel, for example, carries freight packed in large steel boxes, which not only protect the cargo in transit, but are also much easier to handle and stow aboard ship than general cargo. Roll-on, roll-off vessels carry wheeled vehicles that can be towed either on or off the ship or can move under their own power. Vessels in the lighter-aboard ship (LASH) category carry loaded barges, which can be placed in or picked out of the water at ports and towed between ship and shore. "Combo" ships carry both general and wheeled freight, but the vehicles must be lifted on and off the ship in the same manner as ordinary cargo.

Contract carriers are the fabled "tramp" steamers that follow no regular route or sailing schedule. These vessels will pick up and deliver cargo wherever a lucrative shipment is available. A vacation voyage aboard one of these tramp steamers can be an unusual and exciting experience. The trip is slow and relaxed, the accommodations on board are spacious and comfortable for the few passengers carried, and the traveler is taken into ports of call rarely seen on the ordinary ocean excursion.

Charter vessels are hired by a group of several shippers who pool their cargo until they have a shipload. In this way, each shipper saves money on the ocean transportation charge and, if the freight is all going to the same port or in the same general direction, obtains more expeditious transportation.

Shipping conferences are not really types of vessels. A group of vessel owners forms an association, or cartel, each of whom operates on a particular route. Since the rates set by these conferences for their members are generally lower than free-lance operators, they compete directly with nonconference vessel operators. Lower rates, or even rebates, are frequently given to shippers who agree to send their freight only on conference vessels. Were these shipping conferences based in the United States, they would be considered highly illegal rate-fixing conspiracies that violate antitrust and antimonopoly laws. Other nations, however, tend to be more lenient toward cartels; hence these groups of vessel operators continue to be an important element in ocean transportation.

OCEAN FREIGHT RATES

The rate structure in ocean transportation tends to be more complicated than in domestic surface transportation. Overseas shipments often involve a land journey in the countries of origin and destination, separated by an ocean voyage. Carrier liability for loss and damage tends to be less in ocean freight so that more insurance must be taken out by the shipper for full protection of the value of the cargo. The additional handling as the freight changes modes of transportation also puts an especially severe burden on the designers of packaging. Not only must the packaging provide protection against damage and pilferage, but it should be able to withstand the inspection procedures at national boundaries.

Within the United States, the domestic travel of shipments destined for overseas movement is treated differently than for shipments over the same routes not intended for international trade. For example, a shipment traveling overland to a port on one of the Great Lakes may be subject to equalization. Rate equalization is a process in which the domestic rail and water freight rates will be adjusted to keep the total rate equal no matter what lake port is used.

Suppose a shipment originating in Alton, Illinois, could be

shipped through either a port on Lake Erie or a port on Lake Michigan before entering the ocean part of the journey. If the goods go by rail to a Lake Michigan port, a shorter overland distance is traversed; hence, the rail charges will be low but the domestic water rates high. If a Lake Erie port is used, a much longer overland rail haul is involved; hence, the rail charges will be higher, but the domestic water rates low. In this way, the total freight bill, including both the domestic rail and water haul, is kept the same, or roughly so, until the freight enters the channels of foreign trade.

Rates on the same items also will generally be lower for goods destined for overseas than for goods sent to a domestic destination even though both shipments travel the same overland route to the same port area. For example, automobile spare parts shipped from Detroit to the port of Baltimore and destined for shipment overseas will carry a lower rate than the same auto parts consigned to a dealer in Baltimore city.

Another common method of setting ocean freight rates is on a CIF basis. This method includes three elements in the rate:

C = the cost or price of the item shipped.

I = insurance (in this case marine insurance)

F = freight charge

When the rate is set on the CIF basis, title to the goods is transferred to the buyer at the destination port. This makes the buyer responsible for both the transportation and delivery of the merchandise within the destination country. The shipper is thus relieved of dealing with an unfamiliar transportation system and highly changeable government regulations.

The way freight rates are quoted in ocean transportation is so different from the domestic practice that a word of explanation is necessary. First, consider the vocabulary needed to follow the discussion. The size of a shipment is expressed in both weight and volume. The units of weight are the

Short ton = 2,000 pounds

Long ton = 2,240 pounds

The unit of volume is the

Measurement ton = 40 cubic feet

This small example will show the relationship between the units of weight and volume. If a shipment weighs 56 pounds per cubic foot, 40 cubic feet (or 1 measurement ton) of this product will weigh in at

$$56 \text{ pounds} \times 40 \text{ cubic feet} = 2{,}240 \text{ pounds}$$

or one long ton. Therefore, 1 long ton equals 1 measurement ton.

Shipments weighing exactly 56 pounds per cubic foot, however, are not especially common. For most cases, the cargo will weigh in at some other figure, say, 44.8 pounds per cubic foot. One long ton of this product weighing 2,240 pounds would occupy this much cubic footage in the hold of the vessel:

$$\frac{2{,}240 \text{ pounds}}{44.8 \text{ pounds}} = 50 \text{ cubic feet}$$

Suppose this ship line charged $25 per ton for general cargo of the type to be shipped. The ocean freight rate for this cargo when computed by the volume measure would be

$$\frac{\$25}{40 \text{ cubic feet}} = \$0.625 \text{ per cubic foot}$$

The 50 cubic feet in this particular shipment would thus incur a freight charge of

$$50 \text{ cubic feet} \times \$0.625 = \$31.25$$

Were the cost to be computed by weight, the freight charge would be (keeping in mind that 1 long ton of this product is being shipped)

$$2{,}240 \text{ pounds} \times \$25 = \$25.00$$

The ship line is privileged to maximize its revenues by charging either by weight or volume at the option of the carrier. In this instance, the vessel operator would charge by volume and earn $31.25 rather than $25.

Occasionally, a dual rate will be quoted. A quotation of "35/62.5 per 100 pounds" means that the rate will be 35 cents per cubic foot or $62\frac{1}{2}$ cents per 100 pounds, whichever produces the larger revenue. These dual rates are designed to produce the same revenue no matter which way the charge is computed provided the cargo weighs in 40 cubic feet per ton:

40 cubic feet at 35 cents per cubic foot = $14

or the same as 1 long ton, or

22.4 at 62½ cents per 100 pounds = $14

When accepting cargo that does not weigh in at 40 cubic feet per ton, the vessel operator has a choice of computing rates by cubic feet or tonnage, whichever produces the highest revenue.

For many companies, international logistics assumes an importance equal to that of domestic logistics. While many of the functions are the same, some practices are so vastly different as to comprise a distinct specialty within the general field of logistics management. Even firms that do not normally deal in international commerce need some knowledge of international logistics to understand the pricing structure of goods imported from abroad that are in competition with their own domestic manufactures.

We have now traversed a long, sometimes highly technical journey. At this point, so close to the end, we can now enjoy the leisure of waxing a bit philosophical. Since forecasting is far more of an art than a science, and crystal balls tend to be quite murky, the exact shape of the future is difficult to define. But that should not keep us from trying in the next chapter.

OBJECTIVES ACHIEVEMENT CHECKUP

The following exercises will help you in assessing the extent to which you have achieved the objectives for this chapter. Where essay-type answers are required, the more you are able to condense your knowledge into a short paragraph of a few sentences the better. Looking back as you work these exercises and rereading parts of the chapter are permissible; but do not look ahead to the answers until you have done your best with the questions.

1. Describe the relationship between logistics management and international trade.
2. List the similarities and differences between domestic and international logistical functions.
3. Describe the relationship between the value of an inventory and foreign exchange rates.

4. How can the risk of holding inventories abroad be reduced?
5. Match the type of company in column I with its function listed in column II.

Column I: Type of Company	*Column II: Function*	
A. International freight forwarder	(a) Provides special packaging	(a) ___
B. Non-vessel-operating common carrier (NVOCC)	(b) Furnishes credit information	(b) ___
C. Export management company	(c) Books cargo space	(c) ___
	(d) Represents exporters	(d) ___
D. Export packer	(e) Moves freight from shipper to overseas customer	(e) ___
E. Air freight forwarder		
F. Customs house broker	(f) Makes transportation arrangements for importers	(f) ___
	(g) Enters into joint rates without owning any vessels	(g) ___
	(h) Can be part of a surface freight forwarder	(h) ___
	(i) Can serve as the export department of a shipper	(i) ___
	(j) Handles overseas sales	(j) ___

6. Compute the ocean freight charge for the following shipment: The ship line you are going to use charges $35 per ton for the type of freight in your shipment. The goods weigh 32 pounds per cubic foot. Calculate the ocean freight rate by both volume and weight and select the one the ship line will probably use if the shipment weighs in at 3,360 pounds.
7. The company accountant tells you that inventory held in a branch located in Seoul, Korea, is valued at $50,000. The exchange rate today, January 1, 19AA, is

$$\$1 = W50$$

or

$$\$0.02 = W1$$

By the end of the year, current forecasts predict the exchange rate will be W1 = $0.04. What would you do to offset translation gains or losses by the end of the year?

ANSWERS TO CHECKUP QUESTIONS

1. To a significant extent, logistics management makes trade possible, since trade essentially is the movement of goods between areas. All the activities connected with getting goods ready for transport between seller and buyer are logistical in nature. The more highly developed are logistics management and logistical facilities, the more readily can trade occur.
2. a. Similarities:
   (1) Obtain and fill orders
   (2) Store and ship goods
   (3) Select modes of transportation and carrier
   (4) Determine level of customer service
   (5) Prepare shipping documents and records

   b. Differences:
   (1) Less availability of storage facilities
   (2) Frequently, smaller quantities ordered
   (3) Packaging
   (4) Different carrier capabilities
   (5) Preference for goods produced locally
3. The value of an inventory held overseas expressed in terms of the nation in which the goods are located will not be affected by changes in the prices of national currencies. When the inventory must be expressed in terms of the currency of the company's home country, problems develop. The *amount* of inventory does not change when exchange rates fluctuate. When the prices of currencies change, the *value* of the inventory fluctuates.
4. The risk of holding inventory abroad can be reduced by:
   a. Diversification: Buy raw materials and produce in nations whose currencies are stable.
   b. Produce in one country and store the goods in another country.
   c. Reduce the timing of the order cycle and use expedited means of transportation to meet delivery dates.
   d. Increase the timing of the order cycle to keep the logistical pipeline full.
   e. Use economic trade unions or free-trade zones.
5. (a) D (f) F
   (b) C (g) B
   (c) A (h) E
   (d) A (i) A
   (e) B (j) C

6. Computation of the ocean freight charge by volume:

$$\frac{2{,}240 \text{ pounds}}{32 \text{ pounds}} = 70 \text{ cubic feet}$$

$$\frac{\$35}{40 \text{ cubic feet}} = \$0.875 \text{ per cubic foot}$$

$$\text{Freight charge} = 70 \text{ cubic feet} \times \$0.875 = \$61.25$$

Computation of the ocean freight charge by weight:

$$\text{Shipment of } 3{,}360 \text{ pounds} = 1.5 \text{ long tons}$$

$$1.5 \times \$35 = \$52.50$$

Since the carrier is permitted to charge the rate that yields the greatest revenue, the ocean freight charge would be computed by volume and would be $61.25.

7. The value of the inventory in Seoul, Korea, on January 1, 19AA is

$$\frac{\$50{,}000}{\$0.02} = W2{,}500{,}000$$

If the exchange rate forecasts are correct, the value of the inventory on December 31, 19AA, will be

$$W2{,}500{,}000 \times \$0.04 = \$100{,}000$$

for a translation gain of $50,000, which might be taxable in some jurisdictions. To prevent showing a translation gain at the end of the year 19AA, the following could be done:

a. Contract with a foreign exchange specialist to deliver to the dealer on January 1, 19BB, W2,500,000 and receive payment in dollars at the current exchange rate of W1 = $0.02.

b. Then, on January 1, 19BB:
   (1) Buy W2,500,000 for delivery to the currency dealer at the new rate of W1 = $0.04. This will cost your company $100,000.
   (2) Receive payment from the dealer at the contract price of W1 = $0.02 for the *won* currency, or $50,000.

c. The loss of $50,000 on the currency transaction exactly balances the $50,000 gain on the valuation of the inventory.

11

To the Future

CHAPTER OBJECTIVES

Upon completion of this chapter, you should be able to:

A. Differentiate between inventory management and just-in-time inventory control.
B. Identify the requirements for using just-in-time inventory control.
C. Describe possible future changes in the relationship between business and government.
D. List the differences between mechanization and automation.
E. Project into the future logistics management as an integrated system.

What does the future hold in store for business logistics management? Change is certainly inevitable, both radical and gradual. New inventions, new procedures, new techniques are forever being born of the ingenuity of the human mind. Management is best advised to be prepared for change and to overcome the resistance to change found in so many organizations and in the hearts of so many managers. One of the larger dangers is faddism: leaping to embrace a new procedure or adopt a new technique simply because it is new and everyone else seems intent on doing the same. Change is not being resisted, however, if management exercises prudence and pauses to consider the consequences of new ideas that seem attractive at first.

Inventory control is a good example of the process of change. Inventories were badly controlled until World War II, with its vast logistical problems, forced attention to devising and installing better methods. After World War II, perhaps as a legacy of the successful solutions of so many overwhelming logistical problems, business gained much better control over inventories. Materials requirements planning (MRP) became the wave of the future in inventory control in the 1950s and 1960s. Closed-loop MRP, MRP II, distribution requirements planning (DRP), and other variations and extensions of these themes have proved of great value in controlling both operations and costs in the logistical process.

The new wave of the future appears to be kanban, or some of its offshoots such as stockless inventory and just-in-time (JIT) inventory. These cost-conserving methods of controlling inventories are not, despite the enthusiasm surrounding their introduction, logistical panaceas. The logistics manager is thus required to deflate some of the wildly extravagant claims made for these new techniques. We might profit from an examination of this Japanese managerial import.

As we explore kanban, one bit of managerial trivia to keep in mind is that not every Japanese industry from Hokkaido to Kyushu has rushed to adopt the Toyota Company's contribution to logistical management. The lack of resources (Japan's raw materials supply pipeline literally stretches around the world) and space provided a favorable soil in which procedures could be developed aimed at conserving materials and economizing on space. The close relationship in Japan (a relationship close enough to be illegal in the United States) between management, workers and their unions, banks, and the government adds to this favorable climate. Many Japanese companies, however, still operate in a more familiar manner.

JIT/KANBAN INVENTORY CONTROL

Traditionally, manufacturing methods have been divided into two major categories: job lots for items built to individual specifications or fabricated in small batches, and continuous production in which the finished goods are turned out in a steady stream. An electricity-generating turbine would be produced in job lot production, while plate glass or steel tubing would be produced in continuous production. Manufacturing now has a third category: repetitive production in which individual units are produced in large volumes. Large-volume assembly-line production, such as that used in the automobile industry, would fall in this third category.

Materials requirements planning (MRP) has been used successfully to provide parts at the right time to the manufacturing process, taking into consideration lead times, production requirements for finished items, and the requirements of the bills of materials. Repetitive manufacturing, as used in Japanese automobile and electronics industries, has inspired a method of logistical support in which the production shop calls or "pulls" materials into the shop "just in time" to be used. Ideally, the materials are supplied directly from small stocks held in the shop or directly from the suppliers. Thus JIT inventory control requires no safety stocks, no economic order quantities, and virtually no inventory except at the work station, and then only enough to keep the work station functioning.

Just-in-time inventory control with its delivery of parts and materials to work stations offers distinct advantages to users, such as:

- Reduces the frequency of stockouts.
- Reduces the level of inventory.
- Reduces the need for materials-handling equipment.
- Shortens production and delivery times.

These four improvements are but a sampling, because the system, if properly used, does work. The Westinghouse Electric Corporation, for example, reduced stockouts by 95 percent at its Bloomington, Indiana, plant using JIT. To date, a standard name for this method of inventory management has not been adopted by logisticians. In the United States, this system has been called just-in-time (JIT) inventory control or stockless inventory. Even in Japan, where the system supposedly was invented, the procedure goes by various names:

- Toyota calls its system *kanban* (meaning card or ticket).
- Honda terms their system DOPS (daily overhead and perfect supply).
- Nissan uses the term "action plate."

At the risk of some heroic oversimplification, kanban can be described as a visual card control system. As a container of parts or materials is emptied at a work station during production, a control card is returned to signal that a new container of the same parts should be made ready. The effect of this is that parts are ordered only shortly before being used at a particular work station. This procedure requires managers with a great tolerance for stress. The kanban procedure requires an enormous amount of reliability for each element and serves to integrate the components of the logistical process—supplier, manufacturer, and customer—into a single, highly interdependent system. To accomplish this logistical and managerial feat, materials are received into the organization in small amounts and at frequent intervals. The shop, in turn, makes small amounts of the finished product but with considerable frequency. The kanban, or ticket, is used to determine when a small supply of parts is started through the production process and to authorize a re-supply.

The requirements for kanban/JIT are so strict that not every type of manufacturing operation can use the method. When using kanban/JIT:

- Production should be repetitive and in large total volume (frequent model changes or irregular-sized production runs are not well suited to kanban).
- Machine setups should be fairly infrequent and fast.
- Production runs are short and about equal in quantity produced.
- The finished product is fabricated at a series of work stations that can be arranged in sequence as on an assembly line.
- Machine breakdowns that stop the flow of production must be corrected immediately.
- A high level of preventive maintenance is needed.
- Production and work stations should be arranged to use a minimum of floor space to reduce transit time between stations.
- Items at every stage in the fabrication process, from the receipt of raw materials, through manufacture into components, to assembly of the finished product, must arrive at each work station free of defects.

These are difficult conditions to satisfy, but the task is not impossible.

The use of kanban/JIT results in significant savings and improvements in efficiency. For example:

- A 1,000-automobile output in Japan requires 1.5 million square feet of factory floor space; the same output in the United States requires over 2 million square feet.
- The Japanese carry an inventory of $150 worth of materials per auto; American car makers need $775 worth of inventory per auto.
- A hood-stamping operation in an American auto plant requires 6 hours to set up; the Swedes can do the same setup in 4 hours; but the Japanese need only 12 minutes.

The use of kanban/JIT demands faster and more frequent deliveries of parts and materials. Some General Motors plants receive as many as 300 deliveries daily, each delivery providing a small fraction of a day's usage for the particular item. The impact on the transportation and purchasing function of such practices is immediate and severe. Users of kanban/JIT find themselves depending more and more on the greater reliability and flexibility of motor carriers in preference to rail service. Air transportation is used occasionally, but primarily to fly in parts vitally needed to keep the production line going. Some rail carriers, sensing the loss of a significant loss of traffic, are improving their delivery records through various types of expedited service to regain and retain this business.

Suppliers closer to the using plant acquire an advantage in being able to reduce delivery time. This is especially true of suppliers who have their own fleets of trucks and do not have to rely on common motor carriers. In addition, purchasing tends to concentrate its orders on a few suppliers who are able to meet the high delivery requirements and whose products meet "zero defects" standards. Finding suppliers who can meet such rigid delivery and quality standards is no small task. Some vendors can be trained to improve the quality of their service and products. Suppliers who are willing to commit themselves to meeting such specifications will also probably require a long-term commitment from the company—and by "long term" is meant firm contracts extending for more than one year's duration. In some cases, a letter of intent replaces a regular contract, and orders for parts are placed by computer as part of the production planning process. In effect, this turns the vendor into the supply branch of the manufacturer.

Similarly strict requirements are imposed on the production departments as well. Kanban/JIT requires that each daily output target be met without fail. In Japanese factories, workers stay on the job past quitting time, if necessary, to complete the planned output for the day. The two 8-hour shifts common in these factories are separated by 4 hours to provide time for maintenance and overtime work when needed. In fairness to the workers, the shift goes home early if the planned output is reached ahead of schedule. Installing a system like that should make for some interesting negotiations with the unions involved.

Kanban/JIT works equally well with robots on the assembly line since the mechanical workers must still be programmed, scheduled, and supplied. The use of computers becomes especially important when these control techniques are used to ensure that accurate and timely information is being furnished to management.

As helpful as these special control techniques may be, not every industry can use them. A period of intense study is needed to determine whether or not a stockless inventory procedure will work for an individual company. Even if the decision is positive, a conversion to kanban/JIT does not happen overnight. Japanese and American experience indicates that no less than a year is required, even when applied to the most uncomplicated manufacturing processes.

The conclusion is clear, therefore, that kanban/JIT is not just a fad or something to be adopted because it happens to be a "fashionable" management technique. JIT has both rewards and limitations, which need equal recognition. When used properly, the rewards are considerable; when improperly used, it can be a disaster. Thus kanban/JIT takes its place beside existing control techniques without displacing any of them. Curiously, kanban bears a family resemblance to the old Gantt milestone chart, since each job has a time, or milestone, for completion. So we have demonstrated anew that great advances are firmly rooted in what has gone before. Was it not Newton who said that he could see so clearly because he stood on the shoulders of giants?

BUSINESS AND GOVERNMENT

One change in the business environment that has as yet been but dimly perceived is the altered relationship between business and government. Historically, the relationship has been mostly adversarial, with business trying to preserve its freedom to act and government circumscribing this freedom for the benefit of the community. The old comforting ideas that sustained this arrangement are beginning to shift in a new direction.

As an example, consider this question: What is the purpose of a company? The most frequently heard answer is: To make a profit. Certainly, no one is going to maintain that a business should not earn a profit; but other considerations are now taking priority. The answer to the question posed is now becoming: To satisfy the wants and needs of consumers—and then make a profit. After all, how can a profit be made if no one buys the product? Heavy advertising might create a demand for a product, but this is more difficult than in prior years as consumers become more sophisticated and price conscious. But now a larger problem emerges: Who will define the needs and wants of these consumers? And who will direct the efforts of the corporations in meeting those needs and wants?

People, by the way they spend their income, furnish signals to business; but in many cases the signals are blurred and indistinct. Often government must sharpen the focus. Whether this be the federal government defining the needs of the community for clean air, the state government in defining the need to protect the local environment, or the local jurisdiction in defining the needs for honesty in dealing with the community is immaterial. The community, in this world of increasing complexity, ever more frequently falls back on government either to define community needs or to communicate them to business.

Once the needs are defined, how shall the company meet them? Who provides this direction? How about those individuals who own the stock of the company and are, thus, nominally the owners? Hardly. Stockholders in this modern world are merely investors interested primarily in dividends and increases in the value of their shares. A few closely held family corporations in which the owners still hold the majority of the stock are managed as personally owned and operated businesses; but these are an exception. Stockholders do not generally act as owners in providing either direction or supervision of management.

Management is thus left to its own judgment and competence in running the company. Now we are within sight of the core of the new changes in corporate governance. Slowly, reluctantly, clandestinely at times to avoid the unwanted attention of the Antitrust Division of the Department of Justice, a new partnership is being forged between business and government. Business is best suited to provide the competence needed in meeting the needs of the public for goods and services; government is best suited to provide the authority that legitimizes this competence and ability to act.

True enough, this does not sound like the perfectly free market of the economics textbooks in which the self-interest of Adam Smith's "Invisible Hand" leads to desired behavior in the marketplace. In the perfect economic world, the government gives guidance only as a referee to ensure that the rules of the game are followed. The growth of the corporation stemming from the middle of the 1800s and the inexorable trend toward monopoly gives little hope that the free market will ever exist. Government intervention in the economic process has become a way of business life during the last century in an effort (frequently only partially successful) to make economic processes more efficient and equitable. This adversarial relationship has begun to change to a form of partnership, ill defined and only partially understood, but a partnership none the less.

A partnership relation between government and business is not unique. In Japan, economic goals are decided on by a consensus. Suppose the Japanese government, in consultation, decides that Japan will capture the major share of the world semiconductor market. The goal is then approached through close cooperation of the government, which provides the tax breaks and legislative climate (e.g., by, perhaps, authorizing subsidies), the banking system that funds the achievement of the goal, and the managers who will oversee the process to a successful conclusion. Close cooperation of this kind would require a complete restructuring of our thinking about antitrust in the United States if pursued openly and vigorously; but this type of cooperation would transform the current adversarial relationship between government, management, and labor into one of consensus. This transformation may be a long time in coming, or it may be forced on the nation by falling too far behind in the competitive race for world markets. Either way, a change is in the wind, and business, labor, government, and the consuming public had better be prepared for it.

THE TECHNOLOGICAL FUTURE

The effects of technology in the future are just as dimly seen as any other development. What is almost certain, though, is that various applications of automation and mechanization will continue to be made. The best place to begin a discussion of this topic is to distinguish clearly between automation and mechanization:

> MECHANIZATION: Substituting machines for human labor (e.g., stocking warehouse shelves, maintaining accounting records, or preparing purchase orders).

In contrast:

> AUTOMATION: The process of making a system or procedure automatic (e.g., a means of directing a mechanized system to pick stock and transport it to a packaging area without human intervention).

Data processing, which could be called the wave of the present, involves the application of automation to the manipulation of information.

What can automation do for data processing?

- Collect
- Store
- Retrieve
- Display
- Manipulate

What entity does all this? Some extremely smart machines. A *microcomputer* is a small and rather slow computer. Examples of the microcomputer are cash registers that calculate the change due the customer, or the ubiquitous (some teachers of mathematics call it iniquitous because of the damage done to human computing skills) hand-held calculator. The *minicomputer* is a small, compact computer with the higher speeds normally associated with computers but that is limited in the capacity to store data. The *microprocessor* is a minicomputer with a greatly enlarged capacity to store data and control processes. Microprocessors are becoming so cheap and easy to use as technology advances that even small enterprises are finding them within reach financially. They are highly useful to managers.

The repetitive, routine process is especially suited to automation by saving a great deal of time and routine labor. A few of these functions are:

- Display sources of supply, but some human must evaluate the quality of service and product supplied and then select the best source.
- Determine the most economical transportation routing, but some human manager must select the ideal carrier if several carriers serve the same route.
- Check invoices, but some human must approve the payment check sent to the vendor.
- List customers, but some human must evaluate the credit-worthiness and general financial reliability of each customer on the list before accepting an order.
- Control inventory and the flow of materials, but some human must establish the desired level of service and then reconcile machine records with physical reality.

No matter how thoroughly automated a system may be, at some point a human is needed to provide judgment. A steel mill at its automated best might need only a handful of employees on each shift, but the presence of these technicians still demonstrates that the machines cannot do it all themselves. Machines certainly aid the decision-making process, but they cannot make the final decision. This exercise of judgment and the application to the decision of nonquantifiable factors will give managers employment for some time to come.

An example of this kind of interaction between manager and machine might sharpen the focus of this point. The proper keys are pushed on the keyboard and a listing of materials requisitions submitted by the factory appears on the screen. Using the keyboard, a light pencil, or a "mouse" to communicate directly with the computer (voice communication is almost in sight for commercial use), the manager selects one requisition for more detailed analysis. When asked, the machine instantly displays the following information for the selected document:

- How much of the particular item was ordered in the current or previous time periods.
- The vendors who filled the requisition if not in stock in the warehouse.
- The delivery history (e.g., late, on time, early, or in partial shipments).
- Rejections, quantity accepted into stock, and any remaining shortages.

The manager might then direct the machine to furnish further detail concerning any shortages reported:

- When the shop orders were submitted.
- When the purchase order was submitted.
- Vendor response.

At each step of the way, the manager is interpreting the information provided by the machine and using this input in the decision-making process. The machine is not, however, making the decision for the manager.

Other uses of the computer include research by recording and analyzing vendor performance; forecasting by techniques too complex for manual calculation (e.g., linear regression or Box–Jenkins); and project planning and execution [e.g., linear programming, input–output analysis, program evaluation and review technique (PERT), or other methods using higher mathematics]. In these applications, human judgment is often indispensable. For example, forecasts in the late 1970s of greatly increased oil costs simply did not materialize to the extent predicted, mostly for reasons that could not be forecast. The unprecedented willingness of consumers to conserve fuel, followed almost immediately by an oversupply of petroleum on world markets, held prices well under expectations and resulted in forecasts which, at best, had to be modified by a generous ration of common and very human sense.

FUTURE APPLICATIONS

Prophets of the future quickly learn that what is to come is not always simply an extension of what has been or is now. An example is the application of nuclear power in transportation. Here is one wave of the future that got scuttled (to use a highly mixed metaphor). Port captains as a class along with their harbormasters seemed to get unduly nervous whenever a nuclear-powered vessel approached their waters. The civilian populace, in particular, led the expressions of revulsion when the nuclear power plant was installed in a military vessel. To extend the situation, the environmentalists have done a creditable job keeping atomic reactors from providing the power to pull railroad trains.

Objections to nuclear power preclude the development of larger and more efficient power plants for the modes of transportation that would furnish motive power for longer trains and specialized handling equipment to make loading and unloading easier and cheaper. The developments in this area have been fantastic. When the Baltimore and Ohio Railroad was plying its primitive steam trains over the first 14 miles of track in the United States in the early 1830s, sages warned potential passengers that the human body could not withstand the reported speeds of 35 miles per hour. A century and a half later, France, Germany, and Japan, to name a few, operate trains well in excess of 250 miles per hour. A better roadbed to permit attainment of higher speeds and reduce derailments would bring the same speed possibilities to the United States. In the air, the increased use of air freight has stimulated the development of aircraft with more carrying capacity and extended ranges.

If we really want to dream, how about some of these: A passenger- and cargo-carrying submarine would provide a smooth, almost motionless ride, lessening seasickness among the passengers and damage to the cargo. As long as we have plenty of helium gas (which will not burn and explode, as did the hydrogen that destroyed the airship *Hindenberg*) to provide lifting capability, why not return to the dirigible? A judicious selection to avoid destructive weather (which did in most of the zeppelins of history) would again give passengers and cargo a smooth, safe journey at a respectable speed.

Keep a firm grip on your credibility for this one: On land, could we not load passengers and freight in a large pod and push it through a pipeline powered by nuclear energy, solid propellants, compressed air, or perhaps a gigantic spring? Not too long ago the following idea to cheapen ocean transportation was but a dream: Combine sail and steam at sea. Let the winds propel the vessel in fair weather and let the propulsion machinery move the ship when the winds turn foul or the currents contrary. Such a vessel has already been built.

Materials processing certainly can be improved in several ways. Inventory control can aid in reducing carrying costs and speeding order processing. A continuation of the trend toward JIT could serve to integrate suppliers, manufacturers, and customers into a single system. Waste is another area of materials handling that could stand some attention. The current emphasis on environmental protection should generate better ways of disposing of waste materials, hazardous or otherwise, by recycling or reuse.

The trend in governmental regulation is quite blurred. Certainly, the enforcement of safety regulations has become less adversarial in nature and the paper-work burden reduced. Four billion dollars of losses by the major commercial airlines in the first five years of deregulation has spurred some grumbling for reregulation. The fierce price wars occurring after passage of the Airline Deregulation Act of 1978 may slow down the trend toward deregulation in other modes of transportation. One of the most desirable trends would be the emergence of a more coherent government policy toward business. An integrated industrial policy would provide a business environment in which plans could be made without the ad hoc interference of a shelf-full of public agencies applying the patchwork of regulatons now in effect.

One trend that should be started and then pursued vigorously is the management of logistics as an integrated system, rather than as a collection of organizational components loosely stitched together under a corporate name. The loyalty of individuals in purchasing, production, inventory control, transportation and traffic, and engineering to their units is commendable; but the time has arrived (if not passed) to combine this departmental loyalty with the idea that they are first of all working for the company and only secondarily for their respective work units. Any system is composed of units that perform specific and specialized functions. These functions, however, are performed for the benefit and survival of the total organization. Common sense would tell us that when the company does well, the employees also prosper. The recognition of a loyalty higher than the department gives an added increment of flexibility to the system.

Flexibility happens to be a highly desirable aspect of any system. Nothing in logistics ever goes exactly according to plan. Too many variables are involved that could go awry. Materials deliveries are unavoidably delayed, machines inexplicably break down immediately after preventive maintenance, or the computer goes down just when inventory data are needed for a critical managerial decision. The component elements of an integrated system can handle such emergencies and continue to function until the irregularity is corrected. A mere collection of work units or a loose confederation of departments shatters at such times.

Individuals in an organization necessarily must divide their loyalties several ways: to themselves, to their work units, to their departments, and to the company. We could even add to this list loyalty to customers and the entire nation. If each member of each element in the logistical pipeline will work for the benefit of the higher loyalty to the company, the system will be able to function despite occasional delays and breakdowns. In this process, the individual will do as well as the performance of the system. In this way, a true system composed of fully integrated components develops to the advantage of the whole of society.

OBJECTIVES ACHIEVEMENT CHECKUP

The following exercises will help you in assessing the extent to which you have achieved the objectives for this chapter. Where essay-type answers are required, the more you are able to condense your knowledge into a short paragraph of a few sentences the better. Looking back as you work these exercises and rereading parts of the chapter are permissible; but do not look ahead to the answers until you have done your best with the questions.

1. What is the difference between inventory management and JIT inventory control?
2. What are the circumstances under which JIT inventory control can be used best?
3. How are relationships between government and business likely to change in the future?
4. Distinguish between automation and mechanization.
5. What do you foresee in the future for logistics management?

ANSWERS TO CHECKUP QUESTIONS

1. Inventory management includes all activities connected with the management of stocks of materials: ordering, purchasing, receiving, and shipping. Inventory control is an included activity and is concerned with those aspects of inventory management dealing with the flow of stock into and from the inventory, including determining how large the safety stock will be, how much stock will be maintained in the basic stock to meet demand between the time a replenishment order is placed and received, and the size of the reorder quantity. JIT (just in time) is a type of inventory control that brings stock into an organization at the time needed and thus minimizes the stock to be maintained on hand in inventory.

2. JIT inventory control can be used to best advantage when:
   a. Production is repetitive: Individual units (e.g., automobiles) are produced in large volumes.
   b. Machine setup changes or breakdowns are infrequent.
   c. Production runs are short.
   d. Production is carried on at a series of work stations.
   e. Items produced at each stage of the manufacturing process from raw material to finished good are free of defects.
   f. High degree of reliability in the supply of input materials exists.
3. In the future, relations between government and business will probably be less adversarial and more cooperative, with government policy toward business more clearly defined. The internal governance of companies may also become less conflict-prone. A more cooperative arrangement can be expected between labor and management. The stockholders, who are now the legal owners of a corporation, will find themselves playing an even smaller role than they do now. Government can be expected to support the goals of business to improve the competitive position of American firms in international markets.
4. Automation is a process of making a system automatic in operation with a minimum of human intervention. In contrast, mechanization is a process of substituting machines for human labor. Human workers may still operate the machines directly and certainly are needed to provide the link between the individual machine and the logistical process the machine supports.
5. In the future, parochial departmental loyalties will probably be submerged as individuals consider themselves members of a company rather than a work unit team. Improvements in materials-handling equipment and practices will make the logistical system more of an integrated unit than a collection of related functions. This, too, will encourage the idea that individuals are part of their companies and the corporate logistical system, rather than simply employees working for a particular department. As logistical management becomes more effective and efficient, its contribution to the well-being of society should be given greater recognition.

Glossary

ABC classification system: A procedure for assigning inventory items to categories for purposes of control. Can be based on value, annual usage, etc.

Accepting: Permitting a shipment to enter the inventory of a company that has passed inspection and met all other standards.

Air freight forwarder: Consolidates shipments for air transportation to overseas destinations.

All-commodity rate: A rate charged according to the weight of a variety of products loaded into the equipment of a carier, rather than by type of commodity.

Automation: The process of making an entire system automatic (and frequently self-controlled) with a minimum of human intervention.

Basic stock: Items of inventory intended for issue against demands during the resupply lead time.

Bill of lading: A listing of each item in a shipment that:

1. Serves as a receipt to the shipper when the carrier accepts the shipment.

2. Constitutes a contract for the transportation of the shipment.
3. Can also be a certificate of title.

Bill of materials: A list of the components needed to produce an end item.

Blanket order: An arrangement between a buyer and seller specifying a time schedule during which the items ordered will be delivered.

Blanket rate: A special rate charged by a carrier for the movement of a product from one general location to another.

Bonded storage: An arrangement by which fees, duties, taxes, etc., need not be paid until the goods move out of a warehouse.

Break bulk: Dividing a large shipment (frequently a carload) into smaller lots (e.g., less carload or less truckload) at an intermediate point for further movement to several destinations.

Business logistics: All activities required in managing the movement of raw materials to and through production facilities, and finished goods to consumers.

Carriers: An individual transportation company, which can be a member of any of the following categories:

1. Common carrier: Holds its services out to all comers provided it has the authority and necessary equipment to haul the offered products.
2. Contract carrier: Provides transportation services to individual companies under the terms of a contract entered into with the shipper.
3. Private carrier: Owns both the transportation equipment and the freight carried.
4. Exempt carrier: Unregulated carriers who haul specific types of products (e.g., farm and forest products).

Charter vessel: Ship hired by a group of shippers who have pooled the cargo to be carried.

CIF: A procedure for setting ocean freight rates based on:

1. Cost of the item shipped (C).
2. Marine insurance (I).
3. Freight charge (F).

Class rate: A freight rate determined by the type of product and the class into which it is placed for purposes of transportation.

Closed-loop MRP: A broader form of materials requirements planning (MRP), which inclues as an integrated system the master production schedule, the capacity of the plant or factory to produce the item, and the scheduling of items for ordering and receipt.

Combination rate: A freight rate consisting of the sum of all rates charged on a shipment from origin to destination.

Commodity rate: A specific rate charged for the transportation of a named product moving between specific points of origin and destination.

Consumption: The final use of a product.

Coordinating: Ensuring that the component elements of an organization work together in an efficient way.

Critical ratio: A number indicating whether or not a shipment is ahead, on, or behind schedule. Used to determine the order in which shipments are dispatched.

Customer service: All activities concerned with meeting customer needs and wants.

Customs house broker: Performs freight-forwarding functions for importers.

Data processing: Automating the manipulation of information.

Debit memo: An accounting document that informs a supplier that the amount owed (i.e., the account payable on the books of the buyer) is being debited, or reduced, by the cost of the items returned as unacceptable.

Demurrage: A penalty charge imposed when carrier equipment has been held for loading or unloading beyond the allowed free period. The term is used in rail transportation.

Dependent demand: The amount needed for a component item based on the requirements for the end item in which used.

Descriptive requisition: A blank requisition form on which all information must be entered.

Detention: The name of the penalty charge for holding motor carrier equipment beyond the allowed free period. See *demurrage*.

Diminishing returns: The progressively smaller improvement in customer service produced by trying to improve this service through devoting more resources to it.

Directing: Giving instructions and ensuring the availability of resources to support execution of the instructions.

Distribution: Placing raw or finished materials in the hands of the users. Creates time and place utility.

Distribution center: A logistical facility whose primary purpose is the movement of goods into and out of storage as rapidly as possible.

Distribution requirements planning (DRP): A procedure for matching inventory availability with customer (or factory) needs and replenishment of stocks.

Dunnage: The bracing that keeps a shipment from shifting around the interior of carrier equipment.

Economic order quantity (EOQ): A specific number of items in a single order that minimizes the cost of placing the order and carrying the items in stock.

Economic union: An organization of a group of nations designed to facilitate trade among themselves.

Elasticity of demand: A measurement of the responsiveness of volume of traffic carried to changes in rates. It can be:

1. Inelastic: Change in traffic volume is less than proportional to the change in rates.
2. Elastic: Change in traffic volume is greater than proportional to the change in rates.
3. Unitarily elastic: Change in traffic volume is equally proportional to the change in rates.

Exceptions rating: A rating other than the one to which a product is normally assigned.

Exchange: The transfer of goods from wholesale facilities to retail establishments and, eventually, to final consumers.

Export/import rate: A rate applying on shipments with overseas points as origin or destination.

Export management company: An organization that handles overseas sales, including information on the credit-worthiness of potential customers, for exporters.

Export packer: Packages shipments for export to overseas destinations.

Exposure: The risk of gain or loss resulting from holding inventory stocks overseas whose values change as currency exchange rates fluctuate.

Free trade zone: An area in which tariffs, customs fees, and other impediments to the free movement of goods are removed or reduced.

Freight all kinds (FAK) rate: See *all-commodity rate.*

Freight bill: A listing of each item in a shipment with the freight charges.

Form utility: Changing the shape of a product to the desired configuration (e.g., refining crude petroleum into automobile gasoline).

General-purpose warehouse: A public storage facility adapted to receive general types of merchandise and some types that require special treatment and handling while in storage.

Group rate: See *blanket rate.*

Hazardous materials: Substances that if improperly handled or stored are dangerous to human life, equipment, and the environment.

Heuristic programming: A method of reaching an optimal solution to a problem by trial and error.

Incentive rate: A type of volume rate that is designed to encourage shippers to pack more freight in an item of carrier rolling stock (e.g., rail box car, motor carrier trailer) than is specified in a freight tariff as a minimum amount.

Independent demand: Requirements for an item that are not derived from or based on the requirement for other items of which it is a part.

International freight forwarder: A specialist in foreign trade who makes arrangements on behalf of an exporter for the movement of goods to overseas destinations.

Inventory: A stock of goods that is to be used within the organization or that is for sale.

JIT (just in time): A method of inventory control that brings stock into the manufacturing department of an organization just in time to be used, thus requiring little or no inventory except at the work stations.

Joint rate: A combined rate charged on a shipment that moves over the lines of two or more carriers from origin to destination.

Kanban: A method of inventory control pioneered in Japan, and similar to JIT, which brings items to work stations at the time needed rather than passing through an inventory.

Land bridge: A freight routing by which the shipment passes through the United States while moving from a foreign port of origin to a foreign port of destination.

Lawful rate: A freight rate published in a tariff that does not violate some provision of existing law.

Lead-time stock: See *basic stock.*

Legal rate: A freight rate published in a tariff.

Local rate: A freight rate charged on a shipment that moves between points of origin and destination on the lines of only one carrier.

Long ton: A unit of weight measurement consisting of 2,240 pounds.

Make bulk: A process of grouping several small shipments at an intermediate point into a single large shipment for further movement onward.

Master production schedule: A plan listing what, when, and how much of an item will be produced.

Materials management: All activities concerned with the management and control of the flow of raw and semifinished goods from the source of supply to the point of manufacture.

Materials requirements planning (MRP): A scheduling technique for items whose demand is based on requirements for a final product in which the items are used.

Measurement ton: A unit of measurement of volume consisting of 40 cubic feet.

Mechanization: A process of substituting machines for human labor.

Microbridge: A freight routing by which the shipment moves from a foreign port of origin, through a United States port, to an inland city of destination.

Microcomputer: A small, relatively slow computer (e.g., the hand-held calculator).

Microprocessor: A small, compact, high-speed computer with a larger capacity to store and process information and control processes than a minicomputer.

Minibridge: A freight routing by which the shipment moves from a foreign port of origin, through a United States port, to another United States port of destination.

Minicomputer: A small, compact high-speed computer.

Minimum tender: The smallest size of shipment a pipeline will accept for transportation.

Motor carrier: A for-hire highway transportation company:

1. Class I carrier: earns operating revenues of over $5 million annually.
2. Class II carrier: earns operating revenues between $1 and $5 million annually.
3. Class III carrier: earns operating revenues of less than $1 million annually.

MRP II: A practice of incorporating into closed-loop MRP input from finance, marketing, data systems, engineering, sales, and other organizational elements resulting in a single integrated plan. Variously called manufacturing resource planning II, materials resource planning II, manufacturing requirements planning II, etc.

Multiple car rate: A special freight rate charged when several cars can be loaded in the same shipment.

National contract: An order placed on a vendor that consolidates into a single purchase the requirements of several separate plants of one company.

Nonvessel operating common carrier (NVOCC): A common carrier that operates no vessels but consolidates small shipments into larger lots and ensures that the freight moves from the shipper's warehouse to the overseas customer.

Order bill of lading: A shipping document listing the contents of the shipment but which is turned over to the consignee only after the freight charges are paid.

Order cycle: All activities concerned with the receipt, processing, and shipping of customer orders.

Order point/order quantity system: See *reorder point/EOQ system.*

Order point/order-up-to-maximum system: See *reorder point/maximum level system.*

Organizing: Bringing together tasks and human and material resources to achieve company objectives.

Periodic review/order quantity system: See *scheduled review/EOQ system.*

Periodic review/order-up-to-maximum system: See *scheduled review/maximum level system.*

Physical distribution: All activities concerned with the management and control of the flow of finished goods to consumers.

Pipeline: A tube through which products in a liquid or semiliquid state are transported. Petroleum pipelines can be:

1. Gathering lines: Haul crude petroleum from wells to a concentration point.
2. Trunk lines: Haul crude product from a concentration point to refineries.

Place utility: Placing products in the hands of users where desired.

Planning: A process of determining the most efficient way of achieving company objectives.

Private warehouse: A storage facility owned or leased by a firm.

Production: Transforming raw or semifinished goods into final products. Creates form utility.

Protective packaging: Packing that prevents damage to goods while in transit.

Public warehouse: A storage facility that is available to all who wish to use the services offered and is equipped to handle the products tendered for storage.

Purchase order: A formal order to a supplier to provide goods in the specified quantity and quality at the time indicated.

Purchasing: That aspect of materials management concerned with the procurement of materials for use in a company.

Railroad: A for-hire railroad transportation company:

1. Class I carrier: earns operating revenues of over $50 million annually.
2. Class II carrier: earns operating revenues from $10 to $50 million annually.
3. Class III carrier: earns operating revenues of less than $10 million annually.
4. Terminal and switching companies: Performs operations within a rail terminal to switch equipment around and make up trains.

Rate: The money price charged for hauling freight. Usually expressed in cents per hundred pounds (cwt).

Rate basis number: A number that reflects movement between specific points as one element in finding a rate.

Rate differential: The process of determining a rate by adding or subtracting a specific amount (i.e., the differential) from a standard rate.

Rate equalization: Adjusting rates so that the combined rail–water rate through one port is equal to the combined rail–water rate through another port for shipments of freight to overseas destinations.

Rating: A number indicating the class to which a product is assigned as one element in finding a rate.

Receiving: A process of identifying and inspecting inbound shipments of purchased materials to determine the correctness of the quantity and adherence to quality standards.

Released value rate: A special low rate charged when the shipper agrees to reduced carrier liability for loss or damage.

Reorder amount: A quantity of items needed to bring the inventory to a predetermined maximum level. Also called a *reorder quantity (ROQ).*

Reorder point (ROP): A predetermined level of inventory at which a replenishment order should be prepared and submitted.

Reorder point/EOQ system: A method of controlling inventory in which an EOQ is placed to replenish stock when the inventory is drawn down to a predetermined level (the reorder point or ROP).

Reorder point/maximum level system: A method of controlling inventory in which an order is placed to replenish stock to bring the inventory up to a predetermined maximum level when the inventory has been drawn down to a reorder point.

Reparations: Payment by a carrier to a shipper when an unlawfully high rate has been charged.

Requisition: An order submitted by an internal element of an organization for resupply action.

Reserve stock: See *safety stock.*

Safety stock: A quantity of items maintained in inventory to protect against stockouts when demand varies beyond expected limits.

Scheduled review/EOQ system: A method of controlling inventory in which the stock is reviewed at set intervals of time and an EOQ is ordered if replenishment is necesssary.

Scheduled review/maximum level system: A method of controlling inventory in which the stock is reviewed at set intervals of time and replenishment orders are issued that will bring the inventory up to a predetermined maximum level.

Section 22 rates: Special rates charged on government freight. These rates are now authorized by Section 10721 of the Interstate Commerce Act rather than Section 22.

Shipping conference: An association of vessel owners that competes directly with nonconference vessel operators.

Short ton: A unit of weight measurement of 2,000 pounds.

Special-purpose warehouse: A public storage facility capable of handling products that need such special services as temperature and humidity control.

Standard deviation: A measure of the amount of dispersion in a set of data around some central tendency (usually the mean, or arithmetic average).

Stockless purchasing: A method of ordering items (usually at frequent but irregular intervals) in which the products remain the property of the vendor until actually used.

Straight bill of lading: A shipping document that requires the listed items to be delivered only to the named consignee.

Suboptimization: Striving for optimal performance in one element of an organization without regard for the effect this may have on the performance of other elements (e.g., what benefits one department may not provide maximum benefit to the company as a whole).

System contracting: A long-term contract under which the items ordered remain in vendor storage until used.

Tapered rates: Rates that increase as the distance hauled increases but not in the same proportion (e.g., should distance increase by one-third, the rate might increase by one-quarter).

Through rate: A single rate charged a shipper that applies from origin to destination.

Time utility: Placing materials in the hands of users when desired.

Trade zone: An area provided by a national, state, or local government in which fees, duties, and taxes need not be paid until the goods leave the area. While in the zone, goods are on display to potential buyers and some light manufacturing or assembly may be performed.

Traffic management: The control and management of transportation services to include the arrangements for freight and passenger transportation and the use of company-owned transportation equipment.

Trainload rate: A special rate charged when over 20 cars can be loaded at one time to make up an entire train.

Translation loss (or gain): The decrease (or increase) in the value of an inventory held overseas due to changes in foreign exchange rates.

Traveling requisition: A requisition form that contains preprinted item descriptions for products frequently ordered.

Turnover rate: The rapidity with which stocks move in and out of inventory.

Volume rate: A special (and usually lower) rate applied when more than a truckload or carload minimum weight is shipped.

Unit concept: The idea that a shipment should be maintained as a single unit as long as possible within the logistical chain.

Utility: The capability of a product to satisfy a human need or want.

Warehouse: A logistical facility whose primary purpose is storage.

Working stock: Items of inventory intended for issue against demands between reorders.

For Further Reading

Books

BERRY, WILLIAM L., THOMAS E. VOLLMANN, and D. CLAY WHYBARK, *Master Production Scheduling: Principles and Practice* (Washington, D.C.: American Production and Inventory Control Society, 1974).

BLANDING, WARREN, *Blanding's Practical Physical Distribution* (Washington, D.C.: Traffic Service Corporation, 1982).

BROWN, ROBERT G., *Materials Management Systems* (New York: John Wiley & Sons, Inc., 1977).

BUFFA, ELWOOD S., and JEFFREY G. MILLER, *Production-Inventory Systems: Planning and Control* (Homewood, Ill.: Richard D. Irwin, Inc., 1979).

CAVINATO, JOSEPH L., *Finance for Transportation and Logistics Managers* (Washington, D.C.: Traffic Service Corporation, 1980).

EITEMAN, DAVID K., and ARTHUR I. STONEHILL, *Multi-national Business Finance* (Reading, Mass.: Addison-Wesley Publishing Company, 1982).

FOGARTY, DONALD W., and THOMAS R. HOFFMAN, *Production and Inventory Management* (Cincinnati, Ohio: South-Western Publishing Company, 1983).

GROOVER, MIKELL P., *Automation, Production Systems, and Computer Aided Manufacturing* (Englewood Cliffs, N.J.: Prentice-Hall, Inc., 1980).

GUNN, THOMAS G., *Computer Applications in Manufacturing* (New York: Industrial Press, Inc., 1981).

HARPER, DONALD V., *Transportation in America* (Englewood Cliffs, N.J.: Prentice-Hall, Inc., 1982).

JORDAN, HENRY, *Cycle Counting* (Washington, D.C.: American Production and Inventory Control Society, 1976).

LAMBERT, D.M., AND J.R. STOCK, *Strategic Physical Distribution Management* (Homewood, Ill.: Richard D. Irwin, Inc., 1982).

LIEB, ROBERT C., *Transportation: The Domestic System* (3rd ed., Reston, Va.: Reston Publishing Company, 1985).

MAGEE, JOHN R., and D.M. BOODMAN, *Production Planning and Inventory Control* (New York: McGraw-Hill Book Company, 1967).

MARTIN, ANDRE, *DRP—Distribution Resource Planning* (Williston, Vt.: OWL Publications, Inc., 1982).

ORLICKY, JOSEPH, *Materials Requirements Planning* (New York: McGraw-Hill Book Company, 1974).

PLOSSL, GEORGE W., *Manufacturing Control: The Last Frontier for Profits* (Reston, Va.: Reston Publishing Company, 1973).

——, and OLIVER W. WIGHT, *Production and Inventory Control* (Englewood Cliffs, N.J.: Prentice-Hall, Inc., 1967).

SCHARY, PHILIP B., *Logistics Decisions: Text and Cases* (New York: Dryden Press, 1984).

SLETMO, GUNNAR K., and ERNEST W. WILLIAMS, *Liner Conferences in the Container Age* (New York: Columbia University Press, 1981).

TAFF, CHARLES A., *Management of Physical Distribution and Transportation* (Homewood, Ill.: Richard d. Irwin, Inc., 1984).

TERSINE, RICHARD J., *Materials Management and Inventory Systems,* 2nd ed. (New York: Elsevier North-Holland, Inc., 1982).

——, and JOHN H. CAMPBELL, *Modern Materials Handling* (New York: Elsevier North-Holland, Inc., 1977).

WIGHT, OLIVER W., *MRP II: Unlocking America's Productivity Potential* (Boston: CBI Publishing Company, 1981).

——, *Production and Inventory Management in the Computer Age* (Boston: CBI Publishing Company, 1974).

ZENZ, GARY J., *Purchasing and the Management of Materials* (New York: John Wiley & Sons, Inc., 1981).

Periodicals

American Production and Inventory Control Society Conference Proceedings
Aviation Week and Space Technology
Computerworld
Datamation
Distribution
International Journal of Physical Distribution and Materials Management
Interstate Commerce Commission Practitioners' Journal
Journal of Business Logistics
Management Science
Materials Handling Engineering
Modern Materials Handling
Modern Packaging
Production and Inventory Management
Purchasing
Railway Age
Railway Freight Traffic
Traffic Bulletin and Traffic World
Transportation Journal
Waterways Journal

Index